Residential Construction Academy
Facilities Maintenance

Kevin Standiford

THOMSON
DELMAR LEARNING

Australia Brazil Canada Mexico Singapore Spain United Kingdom United States

Residential Construction Academy: Facilities Maintenance
Thomson Delmar Learning

Vice President, Technology and Trades ABU:
David Garza

Director of Learning Solutions:
Sandy Clark

Managing Editor:
Larry Main

Acquisitions Editor:
James Devoe

Product Manager:
John Fisher

Marketing Director:
Deborah Yarnell

Marketing Manager:
Kevin Rivenburg

Marketing Coordinator:
Mark Pierro

Director of Production:
Patty Stephan

Production Manager:
Stacy Masucci

Content Project Manager:
Nicole Stagg

Technology Project Manager:
Kevin Smith

Editorial Assistant:
Tom Best

Cover Image:
© Veer, Inc./Fancy Photography Collection

COPYRIGHT © 2008 Thomson Delmar Learning, a division of Thomson Learning Inc. All rights reserved. The Thomson Learning Inc. logo is a registered trademark used herein under license.

Printed in the United States of America
1 2 3 4 5 XX 08 07

For more information contact
Thomson Delmar Learning
Executive Woods
5 Maxwell Drive, PO Box 8007,
Clifton Park, NY 12065-8007
Or find us on the World Wide
Web at www.delmarlearning.com

ALL RIGHTS RESERVED. No part of this work covered by the copyright hereon may be reproduced in any form or by any means—graphic, electronic, or mechanical, including photocopying, recording, taping, Web distribution, or information storage and retrieval systems—without the written permission of the publisher.

For permission to use material from the text or product, contact us by
Tel.   (800) 730-2214
Fax   (800) 730-2215
www.thomsonrights.com

Library of Congress Cataloging-in-Publication Data:

Standiford, Kevin.
  Residential construction academy facilities maintenance / Kevin Standiford —1st ed.
    p. cm.
  Includes bibliographical references and index.
    ISBN-13: 978-1-4018-6483-5 (alk. paper)
    ISBN-10: 1-4018-6483-X (alk. paper)
  1. Dwellings—Maintenance and repair.
  2. Building management. I. Title.
  TH4817.S73 2008
  690'.24—dc22

2007025496

NOTICE TO THE READER

Publisher does not warrant or guarantee any of the products described herein or perform any independent analysis in connection with any of the product information contained herein. Publisher does not assume, and expressly disclaims, any obligation to obtain and include information other than that provided to it by the manufacturer.

The reader is expressly warned to consider and adopt all safety precautions that might be indicated by the activities herein and to avoid all potential hazards. By following the instructions contained herein, the reader willingly assumes all risks in connection with such instructions.

The publisher makes no representation or warranties of any kind, including but not limited to, the warranties of fitness for particular purpose or merchantability, nor are any such representations implied with respect to the material set forth herein, and the publisher takes no responsibility with respect to such material. The publisher shall not be liable for any special, consequential, or exemplary damages resulting, in whole or part, from the readers' use of, or reliance upon, this material.

Brief Contents

Chapter 1   Customer Service Skills 1

Chapter 2   Organization of Work Tasks 13

Chapter 3   Applied Safety Rules 19

Chapter 4   Fasteners, Tools, and Equipment . . . 41

Chapter 5   Practical Electrical Theory 71

Chapter 6   Electrical Facilities Maintenance . . . 85

Chapter 7   Carpentry 127

Chapter 8   Surface Painting 203

Chapter 9   Plumbing 227

Chapter 10 Heating, Ventilation, and Air-Conditioning Systems 259

Chapter 11 Appliance Repair and Replacement 289

Chapter 12 Trash Compactors 313

Chapter 13 Elevators 319

Chapter 14 Pest Prevention 325

Chapter 15 Groundskeeping 335

Table of Contents

Preface xv

Chapter 1 Customer Service Skills 1
    Objectives 1
    Introduction 1
    An Excellent Customer Service Attitude 1
        Confidence 2
        Competence 2
        Appreciation 2
        Empathy 3
        Honesty 4
        Reliability 5
        Responsiveness 6
        Courtesy 6
        Tone of Voice 7
        Listening 7
        Responding 7
    Avoid the Words "I Can't" 7
    Review Questions 9
    Customer Service Job Sheet 11

Chapter 2 Organization of Work Tasks 13
    Objectives 13
    Introduction 13
    Establish Priority of Work Tasks 13
    Assign Tasks 14
    Carry Out Work Order Systems 15
    Review Questions 16
    Assigning a Task Job Sheet 17

Chapter 3 Applied Safety Rules 19
    Objectives 19
    Introduction 19
    Purpose of OSHA 20
    Basic Fall Protection Safety Procedures 20
    Developing a Fall Protection Work Plan 21

Environmental Protection (EPA) and
Department of Transportation (DOT)
Hazardous Materials Safety Procedures .. 21
   Safety Procedures........................21
   Electrical Safety Procedures 22
   General Safety Precautions.................22
   Safety and Maintenance Procedures for Power
      Tools and Cords 22
   Ladder Safety and Maintenance Procedures 23
   Appropriate Personal Protective Equipment......27
   Safe Methods and Tools for Lifting and Moving
      Materials and Equipment to Prevent Personal
      Injury and Property Damage 28
   Procedures to Prevent and Respond to Fires and
      Other Hazards 29
   Prevent a Fire from Starting in Your Home 31
   Procedure to Prevent Uncontrolled Chemical
      Reactions 31
Review Questions.....................32
Applied Safety Rules Job Sheet #1.........33
Applied Safety Rules Job Sheet #2.........35
Applied Safety Rules Job Sheet #3.........37
Applied Safety Rules Job Sheet #4.........39

Chapter 4   Fasteners, Tools, and Equipment... 41

Objectives 41
Introduction 41
   Fasteners.............................41
   Types of Fasteners.......................41
   Solvents 46
   Hand Tools............................46
   Portable Power Tools 56
Review Questions.....................64
Tools and Equipment Job Sheet #1 65
Tools and Equipment Job Sheet #2 67
Tools and Equipment Job Sheet #3 69

Chapter 5   Practical Electrical Theory 71

Objectives 71
Introduction 71
Basic Electrical Theory 71
Calculate Electrical Load by Using
   Ohm's Law........................77

Difference between AC and DC Currents 78
Correctly Measure Wire Size and Load
     Carry Capacity 79
Emergency Circuits. 79
Emergency Backup Electrical
     Power Systems 80
Review Questions 81
Practical Electrical Theory Job Sheet 83

Chapter 6 Electrical Facilities Maintenance. . . 85

Objectives 85
Introduction 85
Safety, Tools, and Test Equipment. 86
    Safety. 86
    Tools and Test Equipment. 87
Practical Electrical Theory 87
    AC vs. DC. 87
    Single-Phase AC vs. Three-Phase AC 87
    Emergency Backup Systems 88
Electrical Material, Equipment, Fixtures,
    and Devices 88
    NM Cable. 88
Electrical Materials and Supplies 89
    Boxes 89
Procedures 90
    Installing Old-Work Electrical Boxes in a
        Sheetrock Wall or Ceiling 90
Installing a Cut-in Box 91
Electrical Devices and Fixtures 92
    Switches 92
    Receptacles 92
    Fixtures. 94
Procedures 95
    Installing Duplex Receptacles in a Nonmetallic
        Electrical Outlet Box 95
Procedures 97
    Installing Duplex Receptacles in a Metal Electrical
        Outlet Box. 97
Procedures 98
    Installing Feed-Through GFCI and AFCI Duplex
        Receptacles in Nonmetallic Electrical
        Outlet Boxes. 98
Measurement Instruments 100
Procedures. 102
    Using a Voltage Tester. 102

Procedures.................... 104
   Using a Noncontact Voltage Tester.......... 104
Procedures.................... 106
   Using a Clamp-On Ammeter.............. 106
Electrical Maintenance Procedures....... 107
   Troubleshooting..................... 107
Perform Tests.................. 108
   Test Smoke Alarms and Fire Alarms......... 108
   Test GFCI Receptacle................. 109
   Test Medical Alert Systems............. 109
   Replacing Detectors, Devices, Fixtures, and Bulbs. 109
Procedures.................... 114
   Installing Feed-Through GFCI and AFCI Duplex
      Receptacles in Nonmetallic Electrical Outlet
      Boxes........................ 114
Procedures.................... 115
   Installing a Light Fixture Directly to an
      Outlet Box..................... 115
Procedures.................... 116
   Installing a Cable-Connected Fluorescent
      Lighting Fixture Directly to the Ceiling...... 116
Procedures.................... 117
   Installing a Strap on a Lighting Outlet Box
      Lighting Fixture................. 117
Procedures.................... 118
   Installing a Chandelier-Type Light Fixture
      Using the Stud and Strap Connection to a
      Lighting Outlet Box................ 118
Procedures.................... 119
   Installing a Fluorescent Fixture (Troffer) in a
      Dropped Ceiling................. 119
   Test and Replace Fuses............... 122
Review Questions............... 124
Electrical Facilities Maintenance Job Sheet. 125

Chapter 7 Carpentry.................. 127

Objectives.................... 127
Introduction................... 127
   General Properties of Hardwood and Softwood... 127
   Effects of Moisture Content............ 128
   Correctly Identify and Select Engineered
      Products, Panels, and Sheet Goods........ 130
   Framing Components................. 131
   Interior Carpentry Maintenance........... 134
Procedures.................... 135
   Constructing the Grid Ceiling System........ 135

Procedures. 141
   Applying Wall Molding. 141
Procedures. 143
   Applying Door Casings. 143
Procedures. 145
   Hanging Interior Doors 145
Procedures. 154
   Applying Base Moldings. 154
Procedures. 156
   Installing Window Trim 156
Procedures. 159
   Installing Wood Flooring 159
Procedures. 161
   Installing Manufactured Cabinets 161
Installing Cylindrical Locksets 165
   Marking and Boring Holes 165
   Installing the Striker Plate. 166
Procedures. 168
   Cutting and Fitting Gypsum Board. 168
Procedures. 169
   Installing Sheet Paneling 169
Procedures. 172
   Installing Solid Wood Paneling 172
Procedures. 174
   Installing Flexible Insulation 174
Procedures. 176
   Installing Windows 176
   Exterior Carpentry Maintenance 177
Procedures. 177
   Installing Gutters 177
Procedures. 179
   Installing Asphalt Shingles. 179
Procedures. 182
   Installing Roll Roofing. 182
Procedures. 185
   Woven Valley Method. 185
Procedures. 185
   Closed Cut Valley Method. 185
Procedures. 186
   Step Flashing Method 186
Procedures. 187
   Installing Horizontal Siding 187
Procedures. 190
   Installing Vertical Tongue-and-Groove
     Siding. 190

Procedures.......................... 192
   Installing Panel Siding 192
Procedures.......................... 192
   Installing Wood Shingles and Shakes 192
Procedures.......................... 194
   Applying Horizontal Vinyl Siding............ 194
Procedures.......................... 199
   Applying Vertical Vinyl Siding.............. 199
Review Questions 200
Carpentry Job Sheet 201

Chapter 8 Surface Painting **203**
   Objectives........................... 203
   Introduction 203
   Surface Preparation 203
      Identify and Select Proper Surface Finishes..... 204
      Types of Paint 204
      Identify and Select Proper Finishing Tools for Type of Finish 204
      Properly Prepare the Surface for Painting 206
   Fixing Drywall Problems................ 208
      Apply Paint by Using a Roller and a Brush 210
      Painting the Ceiling..................... 210
      Cleaning and Storing of Equipment and Supplies 212
   Review Questions 215
   Painting Job Sheet #1 217
   Painting Job Sheet #2 219
   Painting Job Sheet #3 221
   Painting Job Sheet #4 223
   Painting Job Sheet #5 225

Chapter 9 Plumbing **227**
   Objectives........................... 227
   Introduction 228
   Plumbing Tools....................... 228
      Identify and Select Basic Plumbing Tools 228
      Piping.............................. 230
      Pipe Fittings 232
   Piping Support and Hangers............ 233
      Hangers and Supports 233
   Measuring and Cutting Pipe............ 235
   Unclogging 237
      Using a Toilet Auger to Unclog a Toilet........ 240
      Caulking............................ 240

Applying Caulk................240
Selecting a Caulk..............241
Plumber's Putty...............242
Applying Plumber's Putty........242
Assembling Pipe..................242
Using PVC Cement..............242
Soldering a Copper Fitting onto a Piece of Copper Pipe..................242
Procedures......................243
Joining Plastic Pipe............243
Procedures......................244
Soldering....................244
Plumbing Codes..................248
National and Local Plumbing Codes..........248
Adjusting the Temperature of a Water Heater................248
Recommended Water Temperatures..........248
Testing the Water Temperature.............248
Adjusting the Temperature of an Electric Water Heater...................248
Adjusting the Temperature of a Gas Water Heater 249
Basic Water Heater Replacement........249
Plumbing Leaks..................251
Shower Seals.................251
Repairing a Faucet...............251
Sink Faucet..................251
Review Questions................253
Plumbing Tool Identification Job Sheet #1 . 255
Plumbing Job Sheet #2..............257

Chapter 10 Heating, Ventilation, and Air-Conditioning Systems.........259

Objectives......................259
Introduction....................259
Perform General Furnace Maintenance....260
Tightening Belts..............260
Replacing Belts...............262
Estimating Belt Sizes...........263
Adjusting Pulleys.............264
Replacing Pulleys.............265
Repositioning Pulleys..........266
Replacing Filters on HVAC Units........266
Maintaining the Heat Source on Gas-Fired Furnaces....................267

Perform General Maintenance of a Hot Water or Steam Boiler 268
Perform General Maintenance of an Oil Burner and Boiler 269
Repair and Replace Electrical Devices, Zone Valves, and Circulator Pumps 272
Lighting a Standing Pilot 274
Perform General Maintenance of a Chilled Water System 275
Clean Coils 276
Lubricate Motors 277
Follow Systematic Diagnostic and Troubleshooting Practices 278
Maintain and Service Condensate Systems . 279
Replace Through-the-Wall Air Conditioners . 280
Review Questions 282
HVAC Job Sheet #1 283
HVAC Job Sheet #2 285
HVAC Job Sheet #3 287

Chapter 11 Appliance Repair and Replacement **289**

Objectives 289
Introduction 289
Repair or Replace a Gas Stove 289
   Gas Burner Will Not Light 289
   The Oven Will Not Heat Properly 290
   Replace a Gas Stove 291
   Repair and Replace Electric Stove 293
   Troubleshooting the Problem 293
   Replacing an Element 294
   Replacing a Receptacle 295
   Oven Does Not Heat Properly 296
   Replace the Oven Heating Element 297
   Replace Electric Stove 298
   Troubleshooting an Ice Maker in a Refrigerator . . 298
   Replacing the Ice Maker 298
   Repairing a Refrigerator 298
   Troubleshooting Dishwasher Problems 299
   Dishwasher Overflows 299
   Replace Dishwasher 299
   Repairing a Range Hood 301
   Unclogging the Exhaust Fan 302
   Replacing a Range Hood 303

Repairing Microwaves 303
Troubleshoot Washers 303
Troubleshoot a Dryer 306
Review Questions 308
Appliance Repair and Replacement
Job Sheet #1 309
Appliance Repair and Replacement
Job Sheet #2 311

Chapter 12 Trash Compactors 313

Objectives 313
Introduction 313
General Maintenance 313
Common Trash Compactor Problems
and Solutions 313
Cleaning and Deodorizing 314
General Maintenance of Hydraulic Devices . 315
Perform a Test of the Interlock
Safety Device 315
Check the General Condition of a Dumpster
and Dumpster Area 315
Review Questions 316
Trash Compactor Job Sheet 317

Chapter 13 Elevators 319

Objectives 319
Introduction 319
Check and Inspect Elevator 319
Review Questions 321
Elevators Job Sheet 323

Chapter 14 Pest Prevention 325

Objectives 325
Introduction 325
Nonpesticidal Pest Control 326
Pest Control with Pesticides 326
Effects of Pesticides on Pests 326
Follow Applicable Safety Procedures 331
Review Questions 332
Pest Prevention Job Sheet 333

Chapter 15 Groundskeeping 335

Objectives 335
Introduction 335

Mowing.................................335
Edging..................................336
Mulch...................................337
Using Bushes and Shrubs in Landscapes...338
Planting Bushes and Shrubs.............338
Winterizing the Irrigation System.......338
Manual Drain Method....................338
Automatic Drain Method.................340
"Blow-Out" Method......................341
Spring Irrigation Startup...............342
Starting Up the Irrigation System for Spring....342
Aeration................................343
Pool Maintenance........................344
Cleaning the Pool Deck.................344
Cleaning the Surface of the Pool.......344
Maintaining the Pool...................344
Snow Plowing............................345
Recommended Snow Removal Procedures....345
Small Engine Repair.....................346
Common Problems with Small Engines.....346
Service Recommendations................346
Engine Smokes..........................347
Maintain Public Areas...................348
Recommended Maintenance Procedures for Public Areas.....................348
Repair Asphalt by Using Cold-Patch Material..............................349
Repairing Small Cracks.................349
Repair Large Holes with Cold Patch.....350
Review Questions........................352
Groundskeeping Job Sheet................353

Appendix.................................355
Conversion Tables......................355
Conversion Factors.....................355
Electrical Wire Gauge and Current Chart.......357
Small Engine Recommended Preventive Maintenance Charts....................357

Glossary.................................359

Index....................................375

Preface

About the Residential Construction Academy Series

One of the most pressing problems confronting the building industry today is the shortage of skilled labor. The construction industry must recruit an estimated 200,000 to 250,000 new craft workers each year to meet future needs. This shortage is expected to continue well into the next decade because of projected job growth and a decline in the number of available workers. At the same time, the training of available labor is becoming an increasing concern throughout the country. This lack of training opportunities has resulted in a shortage of 65,000 to 80,000 skilled workers per year. The crisis is affecting all construction trades and is threatening the ability of builders to construct quality homes.

These challenges led to the creation of the innovative *Residential Construction Academy Series*. The *Residential Construction Academy Series* is the perfect way to introduce people of all ages to the building trades while guiding them in the development of essential workplace skills, including carpentry, electrical wiring, HVAC, plumbing, and facilities maintenance. The products and services offered through the *Residential Construction Academy* are the result of cooperative planning and rigorous joint efforts between industry and education. The program was originally conceived by the National Association of Home Builders (NAHB)—the premier association of more than 200,000 member groups in the residential construction industry—and its workforce development arm, the Home Builders Institute (HBI).

For the first time, construction professionals and educators created national standards for the construction trades. In the summer of 2001, NAHB, through the HBI, began the process of developing residential craft standards in five trades: carpentry, electrical wiring, HVAC, plumbing, and facilities maintenance. Groups of carpentry employers from across the country met with an independent research and measurement organization to begin the development of new craft training standards. Care was taken to assure representation of builders and remodelers, residential and light commercial, custom single family and high production builders. The guidelines from the National Skills Standards Board were followed in developing the new standards. In addition, the process met or exceeded American Psychological Association standards for occupational credentialing.

Next, through a partnership between HBI and Delmar Learning, learning materials—textbooks, videos, and instructor's curriculum and teaching tools—were created to teach these standards effectively. A foundational tenet of this series is that students *learn by doing*. Integrated into this colorful, highly illustrated text are Procedure sections designed to help students apply information through hands-on, active application. A constant focus of the *Residential Construction Academy* is teaching the skills needed to be successful in the construction industry and constantly applying the learning to real-world applications.

Perhaps most exciting to learners and industry is the creation of a national registry of students who have successfully completed courses in the *Residential Construction Academy Series*. This registry or transcript service provides an opportunity for easy access for verification of skills and competencies achieved. The registry links construction industry employers and qualified potential employees in an online database facilitating student job searches and the employment of skilled workers.

About This Book

A facility maintenance technician is an individual who is responsible for the day-to-day maintenance and operational tasks that support a commercial facility. Duties often include but are not limited to:

- take responsibility for a variety of activities related to the repair and maintenance of all aspects of the facility
- perform preventive/predictive maintenance per requirements and non-scheduled or emergency maintenance when required to support operations
- perform basic electrical repairs and hard wiring installation of 120/240-volt systems
- make recommendations to modify or replace equipment when necessary to support, demand, or improve building efficiency
- shovel snow and perform other groundskeeping duties as assigned

The *Facility Maintenance* book provides coverage for the areas in residential wiring that are required of an entry-level facility maintenance technician, including the basic hands-on skills and the more advanced theoretical knowledge needed to gain job proficiency. In addition to electrical, other topics such as customer service skills, carpentry, surface painting, plumbing, appliance repair, pest prevention, groundskeeping, and heating, ventilation and air conditioning systems, along with many other aspects of facility maintenance. The format is intended to be easy to learn and easy to teach.

Features of This Edition

This innovative series was designed with input from educators and industry professionals and informed by the curriculum and training objectives established by the standards committee. The following features aid learning:

Learning features such as the Objectives and Introduction set the stage for the coming body of knowledge and help learners identify key concepts and information. These learning features serve as a road map for continuing through the chapter. Learners also may use them as an on-the-job reference.

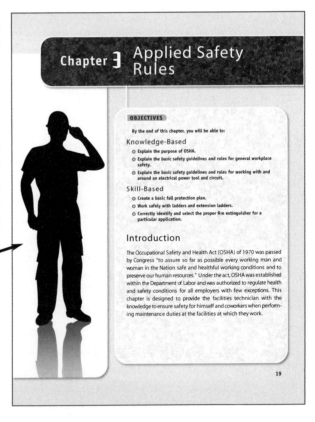

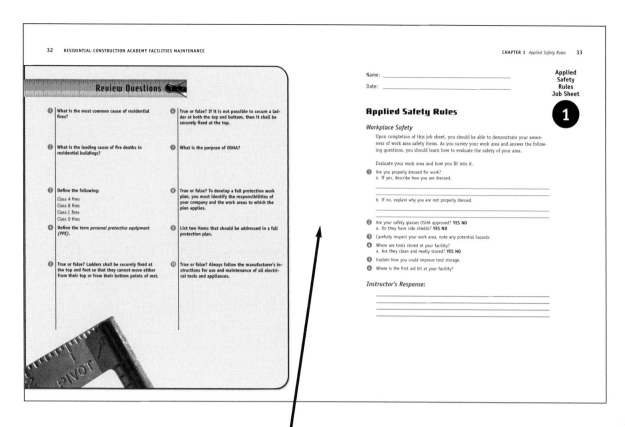

Chapter review questions and job sheets enable the reader to assess the knowledge and skills obtained from reading the chapter.

Step-by-step coverage of typical facilities maintenance tasks explain the work in detail.

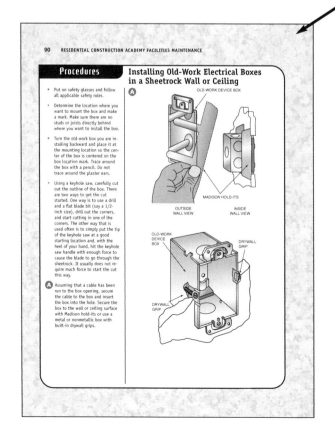

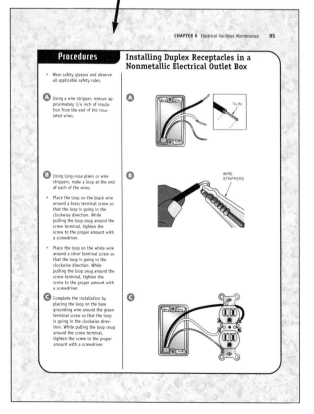

A Glossary containing definitions of terms is an important reference.

In addition, key topic and skill sets are addressed in the following areas:
- Introduction to task management, Chapter 2
- Introduction to customer service skills, Chapter 1
- Introduction to OSHA safety in Chapter 3 Applied Safety Rules, Chapter 3
- Basic concepts of electrical theory, Chapter 5

Turnkey Curriculum and Teaching Material Package

We understand that a text is only one part of a complete, turnkey educational system. We also understand that instructors want to spend their time teaching, not preparing to teach. The *Residential Construction Academy Series* is committed to providing thorough curriculum and preparatory materials to aid instructors and alleviate some of their heavy preparation commitments. An integrated teaching solution is ensured with the text, including the Instructor's e.resource™, a printed Instructor's Resource Guide, and Workbook.

Workbook

Designed to accompany *Residential Construction Academy Facilities Maintenance*, the Workbook is an extension of the core text, and provides additional review questions and problems designed to challenge and reinforce the student's comprehension of the content presented in the core text.

e.resource

Delmar Learning's e.resource is a complete guide to classroom management. The CD-ROM contains syllabi, lesson plans, chapter hints, answers to review questions, and other aids for instructors using this series. Designed as a complete and integrated package, e.resource also provides suggestions for when and how to use the accompanying PowerPoint presentations and test banks. An Instructor's Resource Guide is also available.

Features contained in the e.resource include:
- Instructor Syllabus: goals, topics covered, reading materials, required text, lab materials, grading plan, and terms.
- Student Syllabus: goals, topics covered, required text, grading plans, and terms.
- Lesson Plans: goals, discussion topics, suggested reading, and suggested homework assignments. You have the option of using these lesson plans with your own course information.

- Chapter Hints: objectives and teaching hints that provide direction on how to present the material and coordinate the subject matter with student projects.
- Answers to Review Questions: solutions that enable you to grade and evaluate end-of-chapter tests and exercises.
- PowerPoint presentations: provide the basis for a lecture outline that helps you to present concepts and material. Key points and concepts can be graphically highlighted for student retention.
- Test Questions: over 150 questions of varying levels of difficulty are provided in true/false and multiple-choice formats. These questions can be used to assess student comprehension or can be made available to the student for self-evaluation.

Online Companion

The Online Companion is an excellent supplement for students. It features many useful resources to support the *Facilities Maintenance* book. Linked from the student materials section of www.residentialacademy.com, the Online Companion includes chapter quizzes, an online glossary, product updates, related links, and more.

Acknowledgments

The NAHB and HBI would like to thank the many individuals, members, and companies that participated in the creation of the Facilities Maintenance National Skill Standards. These standards helped guide us in the creation of this text. Thanks also to Quantum Integrations, who helped create the original content, and to Kevin Standiford, who reviewed the material for technical accuracy and created the final manuscript.

In addition, we thank the following people who provided important feedback on the original manuscript, enabling us to hone the content:

Jim Eichenlaub
Director of Government Services
Builders Association of Metropolitan Pittsburgh
Pittsburgh, PA

Michael Frank
Facilities Maintenance Instructor
Quentin M. Burdick Job Corps
Minot, ND

Mark Martin
Carpentry Instructor
Penobscot Job Corps
Bangor, ME

Daryl Martinez
Facilities Maintenance Instructor
Talking Leaves Job Corps Center
Tahlequah, OK

Chapter 1: Customer Service Skills

OBJECTIVES

By the end of this chapter, you will be able to:

Knowledge-Based

- List the attributes of great service.
- Identify personal strategies for connecting with the people you are providing service to.
- Explain the importance of understanding the needs of the people you are providing services to and their expectations.

Introduction

Great service starts with a great attitude about your job, yourself, and the people you are providing service to. The purpose of this chapter is to highlight some important characteristics that lead to an excellent customer service attitude. The facilities maintenance technician will also be introduced to the importance of treating work projects as team projects and exploring the many characteristics required to be an effective team member. Excellent customer service allows service workers to meet the needs of the persons they are providing service to in the most pleasant and efficient way possible.

An Excellent Customer Service Attitude

Attitude really is *everything* when providing service. Attitude sets the stage for all other actions. A positive and service-oriented attitude can help you overcome many problems, dissatisfactions, or mistakes. It can also help in building long-term and loyal relationships with coworkers and customers.

Attitude is also a key factor in first impressions. It tells how you feel about yourself, how you feel about your job, and how you feel about the customer. It reveals confidence in doing an excellent job or it reflects a sense of incompetence.

Finally, attitude is contagious, not only to other employees or contractors with whom you work, but also to the customers themselves. Remember that customers can have their own problems (including attitude problems) and you have an opportunity to improve the customer's day with your positive outlook.

Attitude is made up of several elements: **confidence, competence, appreciation, empathy, honesty, reliability, responsiveness, patience, open-mindedness,** and **courtesy.** If presented properly, these elements assure others that you are capable of doing what you say you can do. Let's look at each of these elements independently.

Confidence

Do you project confidence that you can do your job, solve a problem, or find the information necessary to achieve the customer's goals? What do the people you work with think? Do they have confidence in you that you can do the job?

A high level of **confidence** lowers the stress and anxiety in those around you. Lowered stress and anxiety leads to calm and rational thinking, better relationships, and improved patience, which are all key ingredients to a pleasant and productive work environment.

Confidence is fostered by several behaviors:

- Know how to do your job well.
- Know that the services you provide and represent will meet the customer's needs and expectations.
- If you do not know how to do something or do not have sufficient knowledge, learn it and practice your skills often.
- Believe in your own abilities to complete the task, solve the problem, or find needed information.
- Control your self-talk so that you treat yourself (and others) in a positive way.

Competence

Are you skilled in the tasks you are asked to do? Do you have the appropriate knowledge? Are you efficient so that you can do the tasks quickly and with ease? If you have skills that need improvement, do you know where to go for help? How will you improve your **competence** throughout your career? If you were asked to develop new skills, how would you go about doing this?

You must know how to perform the tasks of your job well. You must understand fully the services you are presented with. You must keep current. You must know where to find information quickly and how to effectively communicate that information to others. Being highly competent tends to increase confidence. The more you know about what you are doing, the better service you can offer your customers.

Appreciation

Do you appreciate your customers? How do you show your appreciation to others? Do you appreciate the skills and know-how you have developed in yourself?

Appreciation is a mindset established through self-talk. **Self-talk** is what you say silently to yourself as you go through the day or are faced with difficult situations.

Self-talk includes all the things you are saying to yourself as you work with a customer and complete your tasks. What you are saying to yourself always comes out in some type of behavior to your customers. It might be in a tone of voice, in an action, or in an easily perceived attitude. You choose what you think about your customer.

Negative or condescending self-talk: *She has no clue about all the things this system can do! What an idiot.*

Positive self-talk: *This is a great opportunity to help her learn about all the things this system can do! Part of what she bought was the training; she is paying for it every bit as much as she paid for the TV. We owe it to her. She is going to be so glad she bought it.*

The kind of self-talk you choose is just that—your choice. Positive self-talk sets you up for a positive attitude and will be followed by positive words and actions. Negative self-talk does just the opposite. It sets you up for a negative attitude followed by damaging words and ineffective actions.

Appreciation also includes how you feel about yourself and the efforts you have made to become successful. Self-talk applies here, too. With negative self-talk, you set yourself up for failure. Positive self-talk can provide just the right amount of confidence and motivation to succeed.

Negative self-talk: *I have no clue how to fix this problem. I might as well give up now before I waste any more time.*

Positive self-talk: *I haven't seen this problem before. This will be a good opportunity to test my skills. Let's see what I can find out.*

Once you make a habit of believing in yourself and your abilities, words and actions will follow accordingly. You will see that you actually find ways to succeed.

Empathy

Do you listen to your customers with empathy and truly want to help to meet their needs? How do you show empathy to those around you? What are the effects when you show empathy?

Empathy does not mean agreeing with everything your customer says. Nor does it mean promising to do everything your customer wants. **Empathy** does mean understanding what your customers need and where they are coming from. Empathy is putting yourself in the other person's shoes, so to speak. Whether you agree with them or not, customers do have a right to their opinions and points of view. They also have a right to be heard. Empathy is realizing this and acting and communicating accordingly.

The first step in empathy is to listen carefully to the customer without forming an opinion or making a judgment based on your own point of view. Taking turns to speak is not really listening. It is just being polite. Really listening means that you do whatever you need to understand the customer's concerns, problem, question, and point of view. Really listening to a customer does two important things: 1) it gives you the information you need to solve the problem or meet the need, and 2) it affirms in the customer that he has been heard. This affirmation is critical and will go a long way toward building trust in you. Once the customer trusts you, you then have the opportunity to address the issue or complete your job successfully. You might be able to provide more information to help the customer fix the problem. You may change the customer's viewpoint by showing that he really didn't want what he thought in the first place. You may simplify what was initially perceived as a complicated issue. Basically, you must listen to the other person before you can expect him to listen to you.

Empathy is important in everyday dealings with customers, but it is especially important when the customer is frustrated, angry, or upset. When people are upset, their ability to think clearly or logically is often diminished. They often say irrational things, demand unreasonable responses, or behave inappropriately. Empathy diffuses emotion. When an angry customer experiences empathy from an employee, the typical response is to calm down and to return to rational and reasonable thinking. With rational and reasonable thinking, most problems can be solved satisfactorily.

Some common phrases that show empathy include:

Wow! I would be mad too if the product broke in the first week.
This problem is definitely an inconvenience. Let's see how to fix this as quickly as possible.
I know it is frustrating when there is this much of a delay. Let me explain why our process is so important.

Be specific in your statements and try to avoid being general. Identify the emotion that you think the customer is feeling. Try to also identify the source of the emotion specifically.

A word of warning: If you try to show empathy without sincerity, you only make matters worse. Your customers are not stupid. They will see through your insincerity and feel patronized.

Honesty

Are you honest with your customers and do your words and actions convey honesty? Do you strive to build trusting relationships with your customers? How specifically do you try to be honest with your customers? What are the consequences of dishonesty?

Honesty is an essential element to success whether you are the CEO or the lowest rung of the ladder. And trusting relationships with customers foster customer loyalty and ongoing business.

Customers need to trust that you will be honest with them. They need to trust that you are providing them with the right information. Dishonesty is pretty easy to detect. It comes through clearly in words, tone, and body language, not to mention in customer dissatisfaction eventually when the dishonesty is found out—and dishonesty is *always* found out sooner or later.

To be honest, you must say the truth, follow through on promises, and state that you do not know the answer when appropriate. Customers will quickly determine that you are trustworthy and will continue to give you their business, often even when you cannot fix a problem or meet the immediate need. And they will tell others about how they were treated.

You can get into "honesty" problems in many different ways. The following list reflects only a few examples that you might have experienced yourself.

- Promising that a product will meet a need when you know it won't
- Underestimating a wait time for service when you know it will be longer
- Stating that a product will be available when you know delivery has been delayed
- Underestimating the cost of a repair when you know that it will cost more

There are some basic tips that will help to ensure that the customer sees you as trustworthy:

- Overestimate wait times. If you serve the customer sooner than expected, the customer will be thrilled.
- Never promise anything you are not sure you can deliver. If necessary, tell the customer you need to check your facts with others and get back to them (and then get back with them when you say you will).
- Don't tell a customer what he wants to hear, unless it is the truth. Instead, tell him the facts and work on solving any problems.
- Explain your actions thoroughly and in terms your customer can understand. Sometimes a customer may distrust you simply because he does not have enough information to know that you are doing what you are supposed to be doing or telling you accurate information.
- Present yourself as working on behalf of the customer. For example, address the customer's needs and offer information or make recommendations related to those needs.
- Provide the advantages and disadvantages. Tell the advantages and disadvantages of a product or service option so that the customer can make informed decisions.

Reliability

How do you relate yourself in terms of reliability? Do you *always* do what you say you will do? Can customers depend on you and your products and services to meet their needs? Are you regularly late, or do you typically run on time? How important is being punctual to you?

Reliability is essential to your customers. Customers want services they can depend on. Basically, reliability means doing what you say you will do and when you will do it. If you have scheduled a service call for 9 AM, then show up at 9 AM (not before, unless you call to ensure it is convenient, and not after). Obviously, things happen that can be out of your control. In these cases, it is important to notify customers accordingly. It is also important to do everything in your power to control the situations. Good planning, accurate estimating of a project's time needs, accurate estimating of travel time, and organization all help in this control.

Reliability is also reflected in availability, prompt replies, quick follow-up, and fast work. Reliability implies accountability for actions and any potential problems. In order to be accountable, you must be available. Can customers contact you? Can your coworkers or office contact you easily? If a message must be left, do you respond quickly?

Keys to reliability are listed here. Add your own keys as they relate to your specific job.

- Ensure that customers and coworkers can contact you.
- If a message must be left, check your messages often and respond immediately.
- If you are out of touch for a specific time, let people know so they won't be disappointed by your lack of response.
- Use an alarm to remind you of important meetings (some watches and cell phones have this feature).
- Show up on time—not before or after the agreed upon time, unless it has been approved ahead of time.

- Learn to accurately estimate how long it will take to complete a task, receive a part, schedule an appointment, etc. Don't guess; instead, wait until you have all the information before giving a time estimation.
- Learn how to do the job right to begin with. Reliability also means being able to depend on the quality of work.

Responsiveness

Do you respond to your customers quickly, accurately, and with the goal of meeting their needs and answering their questions? How do you show your customers that you are responsive? How does responsiveness impact your ability to provide excellent service?

There is nothing more frustrating than being passed from one person to another without getting what you need. Yet, this is often the experience of customers in many businesses. For good service, responsiveness is on the top of the list of key elements.

Courtesy

What does it mean when someone is courteous? Are you courteous when you deal with others? How specifically do you demonstrate courtesy?

Behaving courteously sends a positive and powerful message to customers, whether they are your external paying customers or those internal customers with whom you work on a daily basis. Courtesy is also a habit that once formed, becomes second nature.

Characteristics of a courteous employee are reflected in the following behaviors:

- Saying "Please," "Thank you," and "You're welcome"
- Responding with "Yes ma'am" or "No sir"
- Saying "I'm sorry" or "Excuse me"
- Addressing people by their names and using Mr., Ms., or Miss as appropriate (for example, if you do not know them well)
- Saying "Yes" instead of "Yeah"
- Being friendly
- Smiling often
- Opening doors and allowing others to go through first
- Introducing yourself to new people
- Being attentive and focusing on the person in front of you without being distracted
- Using appropriate language

Courtesy also implies sincerity. Show that you sincerely appreciate your customers by thanking them for their business and being specific.

For example: *Thank you for buying our entertainment system. I know you are going to love it. We really appreciate your business.*

Remember your customers' names because you honestly feel that they are important enough to do so. Be sincere in all of your actions. Your sincerity—or lack of sincerity—will be obvious to your customers.

It is easy for people to forget common courtesies when stressed—for example, when a customer is angry or when you are frustrated because of a problem situation. These are the times when you need to be exceptionally courteous. You should be courteous even when you feel that the person does not deserve it and when customers are not being courteous themselves. This is why developing the habit of being courteous is so important. If courtesy is a habit, you are less likely to forget about it when stressed.

Tone of Voice

Sometimes how you say something means as much as what you say. The wrong tone can cause a misinterpretation of your words. Saying "thank you" in an angry tone serves only to agitate your customer. Asking about the problem in a disinterested tone shows the customer that you are not sincere. Using sarcasm typically causes a customer to become angry and to feel disrespected.

Combining a positive, friendly, and confident tone with positive and confident words such as "Absolutely," "Definitely," "Not a problem," and so forth can be very effective. Say the following phrases with a positive, friendly, and confident tone to get the point:

Absolutely. I can have this fixed in no time!
Definitely. I will order the part for you today.
Not a problem. I will reschedule the service call for that date.
Yes! I will be happy to move this for you.

Try to match your tone to the customer's needs. If the customer is in a hurry, then make your tone urgent and energetic. If the customer is frustrated, use a confident and helpful tone. If your customer is doubtful or has many questions, use a reassuring and confident tone.

Listening

There is a significant difference between listening and hearing. To listen means that you truly attempt to understand what the speaker is saying. Real listening is a highly active process. Without listening, there is not communication—only speaking and hearing.

Effective listening does several things for the relationship:

- It shows that you sincerely care about the customer and the customer's needs.
- It demonstrates attentiveness to the customer.
- It allows you to gain critical information with which to complete your task successfully or to solve the problem.
- For frustrated customers, it reduces irritation by ensuring the customers that they are being heard.
- It fosters an effective and productive relationship between you and the customer.

Responding

How you respond to the customer typically determines how well your service is received, which ultimately translates to either customer satisfaction and loyalty or customer dissatisfaction.

There are several ways to respond effectively to customers that will help to make them feel served and fully attended to.

Avoid the Words "I Can't"

Focus on what you can do rather than what you cannot do. If you cannot do exactly what the customer wants, explain what you *can* do for the customer that either comes close to the customer's request or meets the same need in a different way. Be a problem solver.

To connect with customers, develop your skills in the following strategies and then practice them consistently:

- Evaluate your body language and use it to convey the appropriate messages.
- Focus on how you say things and your tone of voice.
- Be attentive to customers.
- Develop effective listening skills and practice these with customers.
- Respond positively to customers.

Review Questions

Define the following attitudes:

1. **Confidence**

2. **Competence**

3. **Appreciation**

4. **Empathy**

5. **Honesty**

6. **Reliability**

7. **Responsiveness**

8. **Patience**

9. **Open-mindedness**

10. **Courtesy**

Name: _____

Date: _____

Customer Service Job Sheet 1

Customer Service

Customer Service Checklist

Reviewing this checklist will help you continuously improve your customer service skills and keep you focused on the only thing you can control—your own behavior.

1. Put yourself in your client's shoes.
2. Reserve judgment about your client and his problem and listen with an open mind.
3. Listen attentively to everything your customer has to say with genuine interest.
4. Ask questions to clarify your understanding.
5. Tell the client what you can do and why.
6. Be empathetic. This doesn't mean you agree with the person's feelings, but it does indicate you acknowledge them.
7. Make commitments. Commitments guarantee that something will get done. It's also a way to manage the customer's expectations.
8. Meet commitments; don't make a commitment just to get rid of a customer. Make a commitment you can keep.

Instructor's Response:

_____
_____
_____

Chapter 2 Organization of Work Tasks

OBJECTIVES

By the end of this chapter, you will be able to:

Skill-Based

- Establish priority of work tasks.
- Assign tasks.
- Carry out work order systems.

Introduction

To successfully complete a project regardless of its complexity or nature, it must first be divided into tasks that can be assigned and the results measured.

Establish Priority of Work Tasks

Traditional wisdom about setting priorities promises you higher productivity and a greater sense of accomplishment. All you have to do is write out a to-do list, prioritize it by order of importance and urgency (using the ubiquitous A, B, and C labels), and then tackle it, right? Then why, after a period time, is that same C item still on your to-do list? And why, after a busy day of completing tasks, do you still find yourself saying, "I didn't get anything done today"?

There are two reasons that may cause this:

1. Priorities changed during the day, but for good reason. You may not have accomplished A, B, or C on your to-do list, but you did respond appropriately to the additional tasks that you were presented with that day.
2. Don't fall into the "ACT, then THINK" method of setting priorities. To prevent yourself from falling into this method of setting priorities, understand three common priority-setting traps and how you can avoid them.

a. Whatever hits first: Do you "choose" your priorities simply by responding to things as they happen? If so, your priorities are really choosing you. Think about how this general lack of control over your day contributes to your stress level. You need to clarify your priorities by determining each task's importance and level of urgency (that is, "THINK, then ACT"). This means negotiating with people to respond in a time frame that's convenient to you and agreeable to them.
b. Path of least resistance: When was the last time you heard yourself say, "It's just easier to do it myself"? This is not always an incorrect assumption, but if you're saying it too often, you're probably not giving the other workers enough credit or you have the wrong person working for you. Ask yourself these questions: Am I trying to avoid conflict? Does the task at hand require more expertise than the other workers have? Should time or money be invested to train someone to take on some of the lower-priority tasks I am currently performing? Answers to these questions will help you determine what alternative action you need to take.
c. Squeaky wheel: In most situations, it is not hard to identify who the squeaky wheels are. Their requests are always urgent and need to be done right away. Usually, you do the work on their time frame. Unless the request is really urgent, give them a specific time or date when they can expect you to complete the task. Eventually they'll understand that their request to complete tasks need to be prioritized with all of your own prioritized tasks.

Assign Tasks

Assigning tasks isn't just a matter of telling someone else what to do. There is a wide range of responsibilities that you can assign to a person along with a task. The more experienced and reliable the person is, the more unsupervised tasks that can be assigned to the person. The more critical the task, the more cautious you need to be when assigning tasks. It is important that each worker understands his or her part in a job and can perform the assigned task. If a worker does not have the ability to complete an assigned task, as a supervisor or manager, you would have to assign that task to someone who has the ability to complete the task.

Before assigning work to a worker, consider these aspects of the job they will do:

- What hazards are in the workplace environment around the worker?
- Are there special work situations that come up that could lead to new risks for this worker? For example: Are there risks that might be encountered outside the normal work area? Just once a week? During one task to fetch materials?
- Are there occasional risks from coworkers, such as welding or machining, that could affect the workers nearby?
- In slow periods, workers might be asked to "help out" other employees. Ensure that any hazards associated with those jobs are reviewed with the worker, by both you and the coworker who will supervise those tasks.

Ensure that you communicate with the worker about the job tasks clearly and frequently, repeating and confirming this training over the first few weeks of work.

Some workers are overwhelmed with instructions at first, and may need to hear this information repeated more than once. Also:

- Inform workers not to perform any task until they have been properly trained.
- Inform workers that if they don't know or if they are unsure of something, to ask someone first. Get them to think in a safety-minded way about all their work.

Carry Out Work Order Systems

Once tasks have been assigned to the appropriate workers, the tasks will need to be completed. Following are suggestions on how to document the completed tasks.

1. Develop a progress report for the current week—This report should contain the following information:
   - Work accomplished: Document the tasks accomplished during the previous week. The tasks should be very specific in nature.
   - Major findings: Document any issues that were encountered when a specific task was dealt with.
   - Worker: In the documentation, record the name of the person who completed the specific task.
   - Estimated hours to complete: This is used to identify the worker's effectiveness in task estimation. The goal is to compare the predicted value (estimated the week before) with the actual number of hours it took for the team to accomplish a specific task.
   - Actual hours to complete: Record the actual number of hours it took for the worker to complete a specific task.
2. Planning for next week—Plan items for the following week. The report should contain the following information:
   - Work item for next week: Used to list all the tasks that the worker plans to accomplish for the following week.
   - Who: List the worker who is responsible for completing the listed task.
   - Estimated hours to complete: For planning purposes, the estimated hours to completion should be identified to show how long it will take to actually complete the task.

Review Questions

1. Why is it important to assign priority to work tasks?

2. How is priority assigned to a task?

3. List four aspects that should be considered before a task is assigned.

4. Why is it important to estimate the amount of time to complete a task before starting that task?

5. When the amount of time to complete a task spans across multiple days or weeks, why is it important to plan the event of consecutive days?

Instructor's Response:

___

___

___

___

Name: _____

Date: _____

Assigning a Task Job Sheet

Assigning a Task

Assigning a Task Checklist

Reviewing this checklist will help you continuously improve your skill in assigning a task.

Consideration	Yes/No	Comment
What hazards are in the workplace environment around the worker?		
Are there special work situations that come up that could lead to new risks for this worker? For example: Are there risks that might be encountered outside the normal work area? Just once a week? During one task to fetch materials?		
Are there occasional risks from coworkers, such as welding or machining, that could affect the workers nearby?		
In slow periods, workers might be asked to "help out" other employees. Ensure that any hazards associated with those jobs are reviewed with the worker, by both you and the coworker who will supervise those tasks.		

Instructor's Response:

_____
_____
_____

Chapter 3 Applied Safety Rules

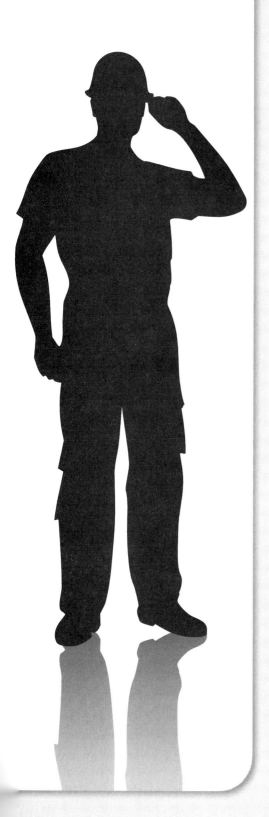

OBJECTIVES

By the end of this chapter, you will be able to:

Knowledge-Based

- Explain the purpose of OSHA.
- Explain the basic safety guidelines and rules for general workplace safety.
- Explain the basic safety guidelines and rules for working with and around an electrical power tool and circuit.

Skill-Based

- Create a basic fall protection plan.
- Work safely with ladders and extension ladders.
- Correctly identify and select the proper fire extinguisher for a particular application.

Introduction

The Occupational Safety and Health Act (OSHA) of 1970 was passed by Congress "to assure so far as possible every working man and woman in the Nation safe and healthful working conditions and to preserve our human resources." Under the act, OSHA was established within the Department of Labor and was authorized to regulate health and safety conditions for all employers with few exceptions. This chapter is designed to provide the facilities technician with the knowledge to ensure safety for himself and coworkers when performing maintenance duties at the facilities at which they work.

Purpose of OSHA

OSHA was created to:

- encourage employers and employees to reduce workplace hazards and to implement new or improve existing safety and health standards;
- provide for research in occupational safety and health and develop innovative ways of dealing with occupational safety and health problems;
- establish "separate but dependent responsibilities and rights" for employers and employees for the achievement of better safety and health conditions;
- maintain a reporting and recordkeeping system to monitor job-related injuries and illnesses;
- establish training programs to increase the number and competence of occupational safety and health personnel; and,
- develop mandatory job safety and health standards and enforce them effectively.

Basic Fall Protection Safety Procedures

Maintaining written fall protection procedures not only protects workers from falls but also protects management from charges of incompetence. Having individual workers or supervisors decide when fall protection is required and what kinds of fall protection equipment to use is an acceptable practice only where workers are routinely exposed to simple hazards, like homebuilders on a roof, for example. However, when workers are involved with lots of non-routine jobs, safety is enhanced if management puts the fall protection and rescue procedures that employees are required to use in writing.

The written plan must describe how workers will be protected when working 10 feet or more above the ground, other work surfaces, or water.

The plan should:

1. Identify all fall hazards in the work area.
2. Describe the method of fall arrest or fall restraint to be provided.
3. Outline the correct procedures for the assembly, maintenance, inspection, and disassembly of the fall protection system to be used.
4. Explain the method of providing overhead protection for workers who may be in or pass through the area below the work site.
5. Communicate the method for prompt, safe removal of injured workers.

Before a fall protection plan can be developed, two important definitions must be understood:

- **Fall arrest system**—equipment that protects someone from falling more than 6 feet or from striking a lower object in the event of a fall, whichever distance is less. This equipment includes approved full-body harnesses and lanyards properly secured to anchorage points or to lifelines, safety nets, or catch platforms.
- **Fall restraint system**—apparatus that keeps a person from reaching a fall point; for example, it allows someone to work up to the edge of a roof but not fall. This equipment includes standard guardrails, a warning line system, a warning line and monitor system, and approved safety belts (or harnesses) and lanyards attached to secure anchorage points.

Developing a Fall Protection Work Plan

To develop a fall protection work plan, you must identify the responsibilities of your company and the work areas to which the plan applies. This information should be listed as the first item in your plan. After you identify your company responsibilities and work areas in the plan:

1. Identify all fall hazards in the work area. To determine fall hazards, you must review all jobs and tasks to be done. After all fall hazards have been identified, list those requiring employees to work 10 feet or more above the ground, other work surface, or water.
2. Determine the method of fall arrest or fall restraint to be provided for each job and task to be done that is 10 feet or more above the ground, another work surface, or water.
3. Describe the procedures for assembly, maintenance, inspection, and disassembly of the fall protection system to be used.
4. Describe the correct procedures for handling, storage, and security of tools and materials.
5. Describe the method of providing overhead protection for workers who may be in or pass through the area below the work site.
6. Describe the method for prompt, safe removal of injured workers.
7. Identify where a copy of this plan will be posted.
8. Train and instruct all personnel in all of the above items.
9. Keep a record of employee training and maintain it on the job.

Environmental Protection (EPA) and Department of Transportation (DOT) Hazardous Materials Safety Procedures

Hazardous materials or chemicals are those substances regulated by federal, state, and local laws, regulations, and ordinances.

Safety Procedures

- Make sure that the names on container labels match the substance names on the corresponding material safety data sheets (MSDSs). If a label is missing or the MSDS is unavailable, notify your supervisor; do not use the chemical until the correct MSDS is obtained. Never remove a manufacturer-affixed label from any container.
- The facilities maintenance technician should be familiar with the hazards associated with the chemicals intended to be used and ensure that all required hazard controls are in place.
- Handle and store hazardous materials only in the areas designated by your supervisor.
- Use an appropriate fume hood or other containment device for procedures that could generate aerosols, gases, or vapors containing hazardous substances.

- When working with highly toxic materials or materials of unknown toxicity, remain in visual and auditory contact with a second person who understands the work being performed and all pertinent emergency procedures.
- Avoid skin contact by wearing gloves, long sleeves, and other protective apparel as appropriate. Upon leaving the work area, remove any protective apparel, place it in an appropriate labeled container, and thoroughly wash your hands, forearms, face, and neck.
- Be prepared for accidents and spills. If a major spill occurs, evacuate the area and dial 911.

Electrical Safety Procedures

Electrical accidents can occur when electricity is present in faulty wiring and equipment or when poor work practices are followed. Accidents involving electricity can lead to burns and tissue damage and in some cases, cardiac arrest and death when the body forms part of the electric circuit. Electric shock can be very unsettling to the victim even if there is no apparent injury.

Other possible consequences of electrical accidents are fire and explosion (as sparking can be a source of ignition) and damage to equipment. Many of the accidents can be traced back to faults such as frayed or broken insulation or practices such as inappropriate work on live equipment.

General Safety Precautions

- Never work on "hot" or energized equipment unless it is necessary to conduct equipment troubleshooting.
- Do not connect too many pieces of equipment to the same circuit or outlet, as the circuit or outlet could become overloaded.
- Be sure that ground-fault circuit interrupters (GFCI) are used in high-risk areas such as wet locations. (GFCIs are designed to shut off electrical power within as little as $1/40$ of a second.)
- Test the meter on a known live circuit to make sure the meter is operating.
- Test the circuit that is to become the de-energized circuit with the meter.
- Inspect all equipment periodically for defects or damage.
- Replace all cords that are worn, frayed, abraded, corroded, or otherwise damaged.
- Always follow the manufacturer's instructions for use and maintenance of all electrical tools and appliances.
- Keep equipment operating instructions on file.
- Always unplug electrical appliances before attempting any repair or maintenance.
- All electrical equipment used on campus should be UL or FM approved.
- Keep cords out of the way of foot traffic so they don't become tripping hazards or become damaged by traffic.
- Never use electrical equipment in wet areas or run cords across wet floors.

Safety and Maintenance Procedures for Power Tools and Cords

Hand and power tools are a common part of our everyday lives and are present in nearly every industry. These tools help us to easily perform tasks that otherwise would be difficult or impossible. However, these simple tools can be hazardous and have the potential for causing severe injuries when used or maintained improperly. Special attention toward hand and power tool safety is necessary in order to reduce or eliminate these hazards.

- Do not use electric-powered tools in damp or wet locations.
- Keep guards in place, in working order, and properly adjusted. Safety guards must never be removed when the tool is being used.
- Avoid accidental starting. Do not hold a finger on the switch button while carrying a power tool.
- Safety switches must be kept in working order and must not be modified. If you feel it necessary to modify a safety switch for a job you're doing, use another tool.
- Work areas should have adequate lighting and be free of clutter.
- Observers should remain a safe distance away from the work area.
- Be sure to keep good footing and maintain good balance.
- Do not wear loose clothing, ties, or jewelry when operating tools.
- Wear appropriate gloves and footwear while using tools.

Ladder Safety and Maintenance Procedures

Ladders can be divided into two main types, straight and step. Straight ladders are constructed by placing rungs between two parallel rails. They generally contain safety feet on one end that help prevent the ladder from slipping (see Figure 3-1).

Step ladders are self-supporting, constructed of two sections hinged at the top. The front section has two tails and steps; the rear portion has two rails and braces (see Figure 3-2).

Safe Ladder Placement

- Ladders, including step ladders, shall be placed so that each side rail (or stile) is on a level and firm footing and so that the ladder is rigid, stable, and secure.
- The side rails (or stiles) shall not be supported by boxes, loose bricks, or other loose packing.
- No ladder shall be placed in front of a door opening toward the ladder unless the door is fastened open, locked, or guarded.

Figure 3-1: **Straight ladder**

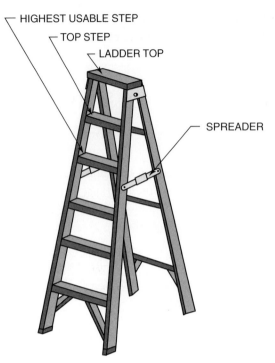

Figure 3-2: **Typical step ladder**

- Straight ladders should be placed against the side of a building or other structure at an angle of approximately 76° (as indicated in Figure 3-3).
- Where a ladder passes through an opening in the floor of a landing place, the opening shall be as small as is reasonably practicable.

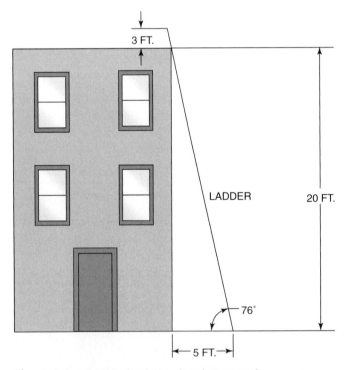

Figure 3-3: **A ladder should be placed at an angle of approximately 76°**

- A ladder placed such that its top end rests against a window frame shall have a board fixed to its top end. The size and position of this board shall ensure that the load to be carried by the ladder is evenly distributed over the window frame.

Safely Securing Ladders

- Ladders shall be securely fixed at the top and foot so that they cannot move either from their top or from their bottom points of rest. If it is not possible to secure a ladder at both the top *and* bottom, then it shall be securely fixed at the base. If this is not possible, then a person should stand at the base of the ladder and secure it manually against slipping.
- Ladders set up in public thoroughfares or other places (where there is potential for accidental collision with them) must be provided with effective means to prevent the displacement of the ladders due to collisions, for example, use of barricades.

Safe Use of Ladders

- Only one person at a time may use or work from a single ladder.
- Always face the ladder when ascending or descending it.
- Carry tools in a tool belt, pouch, or holster, not in your hands, so you can keep hold of the ladder.
- Wear fully enclosed slip resistant footwear when using the ladder.
- Do not climb higher than the third rung from the top of the ladder.
- Ladders made by fastening cleats across a single rail or stile shall not be used.
- When there is significant traffic on ladders used for building work, separate ladders for ascent and descent shall be provided, designated, and used.
- Make sure the weight your ladder is supporting does not exceed its maximum load rating (user plus materials). There should be only one person on the ladder at one time.
- Use a ladder that is the proper length for the job. Proper length is a minimum of 3 feet extending over the roofline or working surface. The three top rungs of a straight, single, or extension ladder should not be stood on.
- Straight, single, or extension ladders should be set up at about a 76° angle.
- Metal ladders will conduct electricity. Use a wooden or fiberglass ladder in the vicinity of power lines or electrical equipment. Do not let a ladder made from any material contact live electric wires.
- Be sure all locks on extension ladders are properly engaged.
- The ground under the ladder should be level and firm. Large flat wooden boards braced under the ladder can level a ladder on uneven ground or soft ground. A good practice is to have a helper hold the bottom of the ladder.
- The two top rungs are not for standing on a step ladder.
- Follow the instruction labels on ladders.

Figure 3-4 and Figure 3-5 illustrate these safety guidelines.

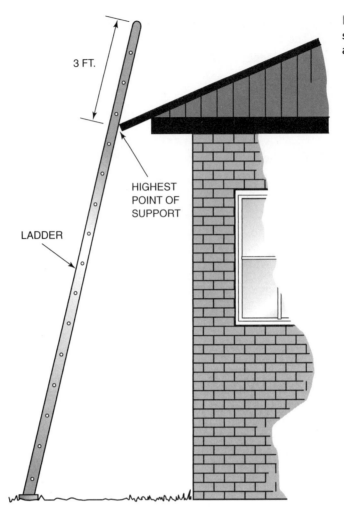

Figure 3-4: The ladders should extend at least 3 feet above the top support

SAFETY PRACTICES FOR STEP LADDERS

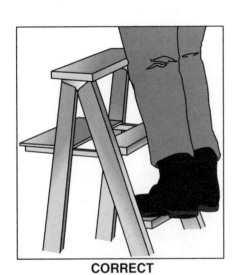

CORRECT INCORRECT

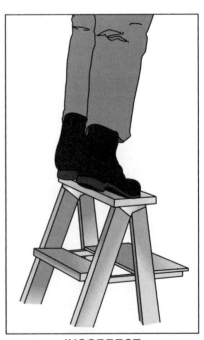

Figure 3-5: Safe practices for using step ladders

Appropriate Personal Protective Equipment

Personal protective equipment (PPE) is defined as all equipment, including clothing for shielding against weather, that is intended to be worn or held by a person at work and that protects him against one or more risks to his health or safety. This equipment includes, but is not limited to, the following:

- Helmets
- Gloves
- Eye protection
- High-visibility clothing
- Safety footwear
- Safety harnesses

To allow the right type of PPE to be chosen, carefully consider the different hazards in the workplace. This will enable you to assess which types of PPE are suitable to protect against the hazard and for the job to be done. Figures 3-6 through 3-11 show examples of PPE.

Figure 3-6: **Typical electrician's hard hat with attached safety goggles**

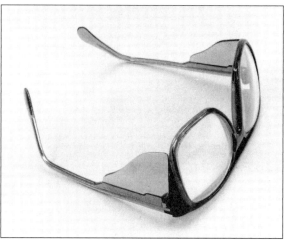

Figure 3-7: **Safety glasses provide side protection**

Figure 3-8: **Leather gloves with rubber inserts**

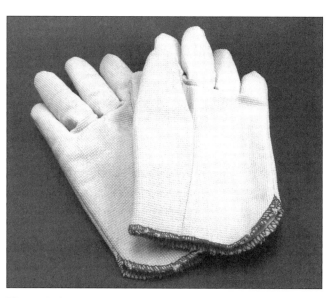

Figure 3-9: **Kevlar gloves protect against cuts**

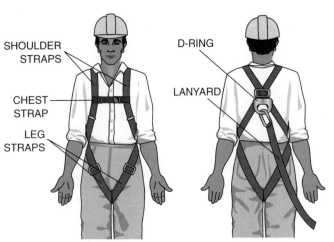

Figure 3-10: Typical safety harness

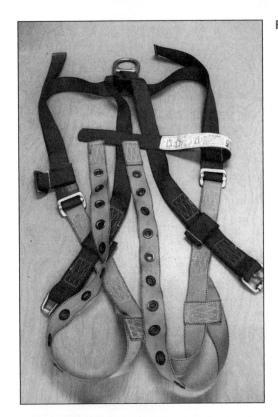

Figure 3-11: Safety harness

Safe Methods and Tools for Lifting and Moving Materials and Equipment to Prevent Personal Injury and Property Damage

General safety principles can help reduce workplace accidents. These include work practices, ergonomic principles, and training and education. Whether moving materials manually or mechanically, employees should be aware of the potential hazards associated with the task at hand and know how to exercise control over their workplaces to minimize the danger.

Proper methods of lifting and handling protect against injury and make work easier. You need to "think" about what you are going to do before bending to pick up an object. Over time, safe lifting technique should become a habit.

Learn the correct way to lift: Get solid footing, stand close to the load, bend your knees, and lift with your legs, not your back (see Figure 3-12).

Figure 3-12: **How to lift safely**

Procedures to Prevent and Respond to Fires and Other Hazards

For a fire to burn, it must have three things: fuel, heat, and oxygen. Fuel is anything that can burn, including materials such as wood, paper, cloth, combustible dusts, and even some metals. Fires are divided into four classes: A, B, C, and D (see Figure 3-13).

Class A Fires

This class involves common combustible materials such as wood or paper. Class A fire extinguishers often use water to extinguish a fire (see Figure 3-14).

Class B Fires

This class involves fuels such as grease, combustible liquids, or gases. Class B fire extinguishers generally employ carbon dioxide (CO_2).

Class C Fires

This class involves energized electrical equipment. Class C fire extinguishers usually use a dry powder to smother the fire.

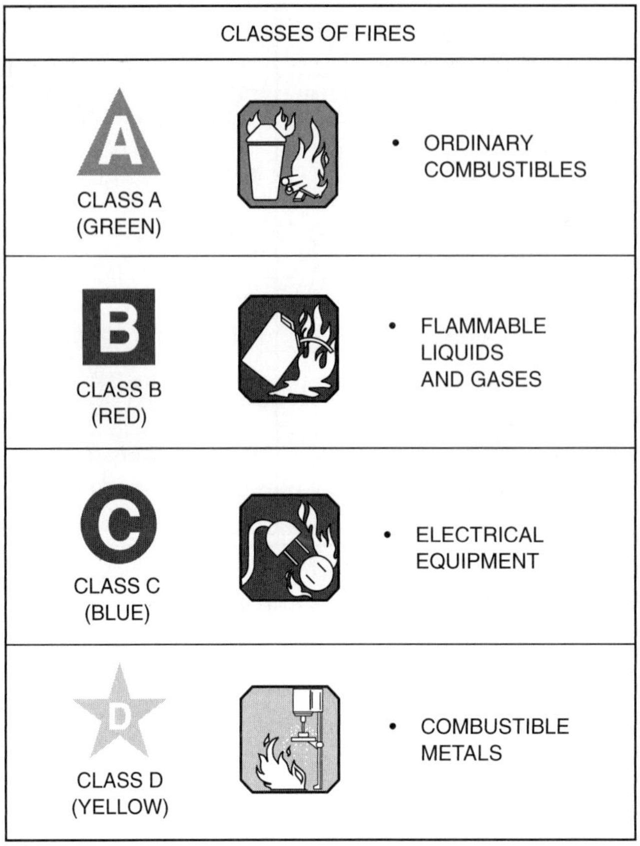

30 RESIDENTIAL CONSTRUCTION ACADEMY FACILITIES MAINTENANCE

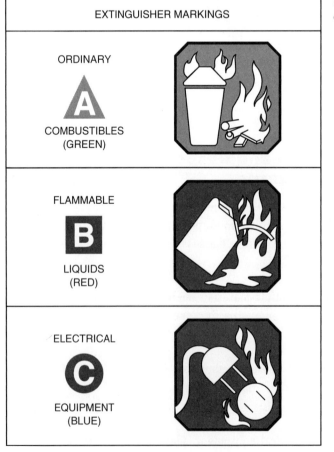

Figure 3-13: Four classes of fires

Figure 3-14: Fire extinguisher symbols

Class D Fires

This class consists of burning metal. Class D extinguishers place a powder on top of the burning metal that forms a crust to cut off the oxygen supply to the metal.

Prevent a Fire from Starting in Your Home

- The most common causes of residential fires are careless cooking and faulty heating equipment. When cooking, never leave food on a stove or in an oven unattended. Avoid wearing clothes with long, loose-fitting sleeves. Have your heating system checked annually, and follow manufacturer's instructions when using portable heaters.
- Smoking is the leading cause of fire deaths and the second-most common cause of residential fires. If you are a smoker, do not smoke in bed, never leave burning cigarettes unattended, do not empty smoldering ashes in the trash, and keep ashtrays away from upholstered furniture and curtains.
- Keep matches and lighters away from children. Safely store flammable substances used throughout the home. Never leave burning candles unattended.

Procedure to Prevent Uncontrolled Chemical Reactions

A chemical reactivity hazard is a situation with the potential for an uncontrolled chemical reaction that can result directly or indirectly in serious harm to people, property, or the environment.

To maintain a safe and healthful working environment, the Department of Energy recommends the following practices wherever chemicals are stored. These practices are based on regulations, rules, and guidelines designed to reduce or eliminate hazardous incidents associated with the improper storage of chemicals.

1. Adhere to the manufacturer's recommendations for each chemical stored, noting any precautions on the label.
2. Label all chemicals. The name and address of the manufacturer or other responsible party must be listed on the label. Chemicals with a shelf life should be labeled with the date received.
3. Store chemicals in the locations recommended (that is, where the temperature range, vibration, or amount of light does not exceed the manufacturer's recommendations).
4. Inspect annually all chemicals in stock and storage. Hazardous chemicals should be inspected every six months. Some hazardous chemicals may require more frequent inspections. Any outdated materials should be properly disposed of or replaced if necessary.
5. Ensure that provisions are made for liaison with local planning committees, the state emergency planning commission, and local fire departments in the event of a chemical emergency.
6. Keep only enough inventory necessary for uninterrupted operation. Chemical inventory should be maintained at a minimum to reduce fire, exposure, and disposal hazards.
7. Rotate new shipments of chemicals with existing stock so that the oldest stock is available first.

Review Questions

1. What is the most common cause of residential fires?

2. What is the leading cause of fire deaths in residential buildings?

3. Define the following:
   Class A fires
   Class B fires
   Class C fires
   Class D fires

4. Define the term *personal protective equipment* (PPE).

5. True or false? Ladders shall be securely fixed at the top and foot so that they cannot move either from their top or from their bottom points of rest.

6. True or false? If it is not possible to secure a ladder at both the top and bottom, then it shall be securely fixed at the top.

7. What is the purpose of OSHA?

8. True or false? To develop a fall protection work plan, you must identify the responsibilities of your company and the work areas to which the plan applies.

9. List two items that should be addressed in a fall protection plan.

10. True or false? Always follow the manufacturer's instructions for use and maintenance of all electrical tools and appliances.

Name: _____

Date: _____

Applied Safety Rules Job Sheet

1

Applied Safety Rules

Workplace Safety

Upon completion of this job sheet, you should be able to demonstrate your awareness of work area safety items. As you survey your work area and answer the following questions, you should learn how to evaluate the safety of your area.

Evaluate your work area and how you fit into it.

1. Are you properly dressed for work?
   a. If yes, describe how you are dressed.

   _____

   _____

   b. If no, explain why you are not properly dressed.

   _____

   _____

2. Are your safety glasses OSHA approved? **YES NO**
   a. Do they have side shields? **YES NO**

3. Carefully inspect your work area, note any potential hazards.

4. Where are tools stored at your facility?
   a. Are they clean and neatly stored? **YES NO**

5. Explain how you could improve tool storage.

6. Where is the first aid kit at your facility?

Instructor's Response:

_____

_____

_____

_____

Name: _____

Date: _____

Applied Safety Rules Job Sheet 2

Applied Safety Rules

Identifying and Handling Hazardous Materials

Upon completion of this job sheet, you should be able to demonstrate your ability to identify hazardous materials and explain how to handle them.

1. Inspect your facility. Identify and list all hazardous materials found.
   a. Solvents
   _____
   _____

   b. Gasoline
   _____
   _____

   c. Cleaners
   _____
   _____

   d. Others
   _____
   _____

2. Check the containers in which hazardous materials are stored. Are they clearly marked? **YES NO**

3. Check to see if your facility has a Materials Safety Data Sheet (MSDS) file. Is it located near the hazardous materials? **YES NO**

4. Make sure your facility has an MSDS list posted on a bulletin board where everyone can read it.

5. Read the MSDS bulletins on each of the materials you find at the facility and explain to the instructor how you would handle a spill of each material.

Instructor's Response:

_____
_____
_____
_____

Name: _____

Date: _____

Applied Safety Rules Job Sheet 3

Applied Safety Rules

Basic Fall Protection Safety

Upon completion of this job sheet, you should be able to demonstrate your understanding of basic fall protection safety procedures.

1. Does your facility have written fall protection and rescue procedures?
   a. If yes, are the procedures readily accessible to the workers?

   _____
   _____

   b. If no, explain why this should be brought to the attention of your supervisor.

   _____
   _____

2. When working higher than 6 feet off the ground, do workers use a fall arrest system?
   a. If no, should this be brought to the attention of your supervisor?

   _____
   _____

3. Identify all fall hazards at your facility.

4. Explain the method of providing overhead protection for workers who may be in or pass through the area below a work site.

   _____
   _____

Instructor's Response:

_____
_____
_____
_____

Name: _____

Date: _____

Applied Safety Rules Job Sheet 4

Applied Safety Rules

Ladder Safety and Maintenance Procedures

Upon completion of this job sheet, you should be able to demonstrate your understanding of ladder safety and maintenance procedures.

1. Name the two main types of ladders and explain the difference between the two.
   _____
   _____

2. Explain why it is important for straight, single, or extension ladders to be set up at about a 76° angle.
   _____
   _____

3. Explain why it is important for the ground under the ladder be level and firm.
   _____
   _____

4. List the ladders at your facility and write down the weight limitations for each.
   _____
   _____

Instructor's Response:

_____
_____
_____
_____

Chapter 4: Fasteners, Tools, and Equipment

OBJECTIVES

By the end of this chapter, you will be able to:

Knowledge-Based

- Describe the safe use of tools, including power tools used by facilities maintenance technicians.
- Describe, select, and install the proper anchors, fasteners, and adhesives necessary for a specific project.
- Select and properly use the appropriate hand tool for a specific project.
- Select and properly use the appropriate power or stationary tool for a specific project.

Introduction

Choosing the appropriate tool or equipment for a project is an important element in being able to maintain any facility. There are many types of tools available for handling many of the problems that may occur. There are times when tools are not the appropriate method for repair, but other equipment may be necessary. A variety of fasteners, solvents, and adhesives will also be introduced in this chapter.

Fasteners

A fastener is a mechanical device that is used to mechanically join two or more mating surfaces or objects.

Types of Fasteners

Fasteners are now on the market for just about anything. They can be used where wood meets wood, concrete, or brick, and most are approved by the Uniform Building Code requirements. However, you should always consult your local building code before selecting a particular type of fastener to incorporate.

Nails

There is a huge difference in nails; some nails are specialty nails.

- **Common nail**—most often nail used. Used for most applications in which the special features of the other nail types are not needed (see Figure 4-1)
- **Box nails**—used for boxes and crates
- **Finishing nails**—can be driven below the surface of the wood and concealed with putty so that it is completely hidden
- **Casing nails**—used for installing exterior doors and windows
- **Duplex nails**—used for temporary structures, such as locally built scaffolds
- **Roofing nails**—used for installing asphalt and fiberglass roofing shingles
- **Masonry nails**—used when nailing into concrete or masonry

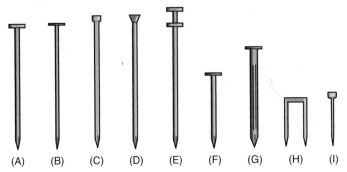

Figure 4-1: Some of the most common kinds of nails: (A) common nail, (B) box nail, (C) finishing nail, (D) casing nail, (E) duplex nail, (F) roofing nail, (G) masonry nail, (H) staple, and (I) brad.

Screws

Screws are used when stronger joining power is needed, or for when other materials must be fastened to wood. The screw is tapered to help draw the wood together as the screw is inserted. Screw heads are usually flat, oval, or round, and each has a specific purpose for final seating and appearance (see Figure 4-2).

Types of screws include the following:

- **Drywall screws**—used to attach drywall to wall studs.
- **Sheet metal screw**—used to fasten metal to wood, metal to metal, plastic, or other materials. Sheet metal screws are threaded completely from the point to the head, and the threads are sharper than those of wood screws. Machine screws are for joining metal parts such as hinges to metal door jambs.
- **Particleboard and deck screws**—corrosion-resistant screws used for installing deck materials and or particleboard.
- **Lag screws**—used for heavy holding and are driven in with a wrench rather than a screwdriver.

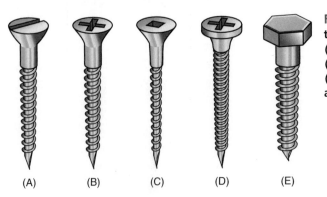

Figure 4-2: Common screw types: (A) wood screw, (B) twinfast drywall screw, (C) particle board screw, (D) panhead sheet metal screw, and (E) lag screw.

Tip: Screw length should penetrate two thirds of the combined thickness of the materials being joined. Use galvanized or other rust-resistant screws where rust could be a problem.

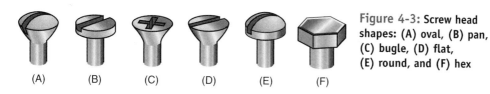

Figure 4-3: Screw head shapes: (A) oval, (B) pan, (C) bugle, (D) flat, (E) round, and (F) hex

Screw head shapes are usually determined by the screw types and the pitch and/or depth of the threads. The most common are oval, pan, bugle, flat, round, and hex (Figure 4-3). There are also different types of slots for these screws (Figure 4-4).

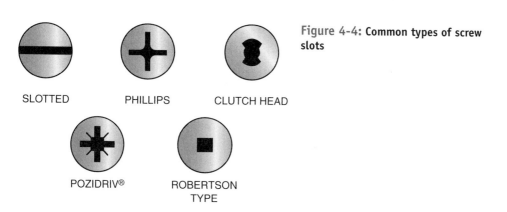

Figure 4-4: Common types of screw slots

Bolts

Nuts and bolts are usually used at the same time: the bolt is inserted through a hole drilled in each item to be fastened together, and then the nut is threaded onto the bolt from the other side and tightened to give a strong connection. Using nuts and bolts also allows for the disassembly of parts.

Types of bolts include the following:

- **Cap screws**—available with hex heads, slotted heads, Phillips head, and Allen drive (Figure 4-5).
- **Stove bolts**—either round or flat heads and threaded all the way to the head. Used to join sheet metal parts (Figure 4-6).
- **Carriage bolts**—a round-headed bolt for timber; threaded along part of the shank; inserted into holes already drilled (Figure 4-7).

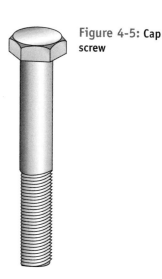

Figure 4-5: Cap screw

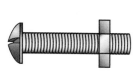

Figure 4-6: Stove bolt

Figure 4-7: Carriage bolt

Adhesives

Sometimes when nails and screws just aren't holding it together by themselves, adhesives are needed.

Types of adhesives include the following:

- Carpenter's wood glue—a white, creamy glue, usually available in convenient plastic bottles. Mainly used for furniture, craft, or woodworking projects, polyvinyl sets in an hour, dries clear, and will not stain. However, it is vulnerable to moisture.
- Epoxy—the only adhesive with a strength greater than the material it bonds. It resists almost anything from water to solvents. Epoxy can be used to fill cavities that would otherwise be difficult to bond. Use it in warm temperatures but read the manufacturer's instructions carefully, since drying times vary and mixing the resin and hardener must be exact.

Tips for using epoxy include:

- One gallon of epoxy will cover: 12.8 square feet at 1/8-inch thickness, 6.4 square feet at 1/4-inch thickness.
- Although epoxies generally are able to withstand high temperatures for short periods we do not recommend using them near or in temperatures above 200°F.
- To achieve maximum adhesion, remove oil, dirt, rust, paint, and water. Use a degreasing solvent (alcohol or acetone) to remove oil and grease. Sand or wipe away paint, dirt, or rust. Roughing up the surface increases surface area for a better bond.
- Epoxy cures quickly enough that there is significant strength after three hours. At 70°F the working time is 15 minutes. It is still possible to reposition work up to 45 minutes after application.
- Uncured epoxy will clean up with soap and water or denatured alcohol. Wash contaminated clothing. Cured epoxy can be removed by scraping, cutting, or removing in layers with a good paint remover.

Contact cements

Contact cement is used to bond veneers or to bond plastic laminates to wood for table tops and counters. Coat both surfaces thinly and allow to dry somewhat before bonding. Align the surfaces perfectly before pressing together, because this adhesive will not pull apart. Use in a well-ventilated area.

Caulk

Caulk is used around outside window and door frames, and to fill outside wall and foundation cracks.

Types of caulk include:

- **Painter's caulk**—inexpensive latex caulk is often used by painters to plug holes and cracks prior to painting. It can also be used to provide a smooth joint in a corner where textured materials meet. This allows the painter to paint a very straight line in the corner when using contrasting paint colors.
- **Acrylic latex**—paintable, acrylic fortified caulk for both interior and exterior applications. Acrylic latex caulk cleans up with water.
- **Siliconized latex**—very durable, latex caulk with silicone. It is available in colors and cleans up with water.
- **100% silicone**—is great for non-porous substances. It is the best choice for sealing ceramic tile, glass, and metal surfaces, but is less appropriate for porous surfaces like wood and masonry. Silicone caulk remains flexible and is impervious

to water. It cannot be painted and must be cleaned up with solvent. It also has a sharp odor when curing. It requires adequate ventilation, and is usually available in either clear or white.
- **Tub and tile**—acrylic sealant gives a flexible, watertight seal. It is mildew resistant with water clean-up.
- **100% silicone kitchen and bath sealant**—has the same characteristics as plain 100% silicone sealant.
- **Gutter and foundation sealant (butyl rubber)**—can be used on metal, wood, or concrete. Appropriate for use in areas that experience extreme temperature variations. Requires solvent clean-up. It is often used on metal flashing and around skylights.
- **Roof repair caulk**—convenient butyl rubber/asphalt formulation for sealing flashing, roofing, skylights, and so on. Cleans up with mineral spirits.
- **Adhesive caulk**—used as an adhesive during the installation of sinks, countertops, and so on. It dries harder than other caulks, and so is less flexible.
- **Concrete and mortar repair**—retains some elasticity to remain in cracks in mortar and concrete. Cleans up with water.

Applying Caulk

Step 1. Caulk comes in either a squeeze tube or as a cartridge (see Figure 4-8). If applying caulk with a cartridge and caulk gun, cut the tapered cartridge nozzle at a 45° angle, with the diameter of the opening equal to the size of the gap. Poke a hole through the tip with a piece of wire to break the internal seal.

Step 2. Apply caulk holding the tube or gun at a 45° angle. Use even pressure to squeeze the tube or trigger (Figure 4-9).

Figure 4-8: **Caulk is made of a variety of materials**

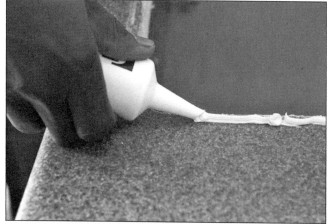

Figure 4-9: **Person applying caulk**

Step 3. Once it begins flowing, move the tip at an even pace along the joint. Remember that less is better. It's easier to apply more caulk than to remove too much.

Step 4. Once you have applied all of your caulk, use a caulk smoother to even out the finish. If you don't have a caulk smoother, then just wet your finger or a popsicle stick to smooth out the bead of caulk.

Step 5. When done applying all of your caulk, give it time to thoroughly set. It should indicate on the package how much time is required for the caulk to set. If you plan to paint over the caulk, it is important that the caulk has completely set before you start painting.

Tip: If you're caulking around your bathtub, do it right. Fill the tub with water before you start. Tubs tend to sink ever so slightly when they are full. So, when you caulk an empty tub, you may not apply enough caulk to compensate for the sinking, which means you'll end up with cracked caulk the next time someone takes a bath.

Tips on bonding:

1. With the exception of epoxy, too much adhesive will weaken the hold of the materials you are bonding.
2. Rough up smooth surfaces slightly before applying adhesives so they will grip more securely.
3. Apply a thin coat of glue, clamp securely, and allow to dry the recommended amount of time.
4. Wipe away excess glue immediately after clamping.

Solvents

Solvents are used in hundreds of products to improve cleaning efficiency. The solvents in cleaners help make counters, showers, toilets, tubs, carpets, and other items clean.

- **All purpose cleaner**—non-hazardous, biodegradable, non-toxic, super-concentrated cleaner
- **Spot remover**—removes chewing gum, tar, grease, make-up, crayon, candle wax, and adhesives from most surfaces
- **Basin, tub, and tile cleaner**—cleans and removes soap scum, lime scale, body oils, and other deposits from shower walls and doors, sinks, ceramic tile, fiberglass, vinyl, porcelain, and stainless steel

Hand Tools

Hand tools, such as hammers, screwdrivers, levels, and tapes do not use a motor.

Measuring, Marking, Leveling, and Layout Tools

Accuracy and care in measuring are all-important. They can mean the difference between a well-put-together project and a sloppy one. Tools that fall in this category are tape measures, squares, combination squares, chalk-line reels, levels, and plumb bobs.

Tape Measures

Tape measures are available in lengths ranging from 6 to 100 feet. A blade that is 1 inch (or more) wide will be safer and easier to use. Those with cushioned bumpers protect the hook of the tape from damage, which is likely to occur when the tape retracts back

CHAPTER 4 *Fasteners, Tools, and Equipment* **47**

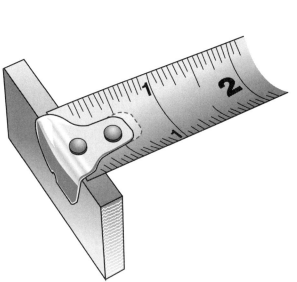

Figure 4-10: **The hook or fitting on the end of a tape measure will slide to adjust for the thickness of the fitting**

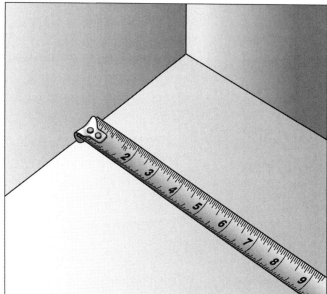

Figure 4-11: **Taking an inside measurement**

into the case. The play in the hook (see Figure 4-10) allows you to make either inside or outside measurements without having to compensate for the hook (see Figure 4-11). Its flexibility allows it to measure round, contours, and other odd-shapes. When making inside measurements, add the measurement of the tape case, usually marked on the case (see Figure 4-12).

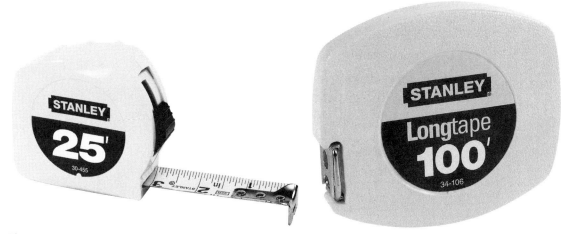

Figure 4-12: **Tape measure**

Squares

Squares are used for laying out work, checking for squareness during assembly, and marking angles. The carpenter's square, also called a framing square, is used for marking true perpendicular lines to be cut on boards and for squaring some corners, among other things (see Figure 4-13).

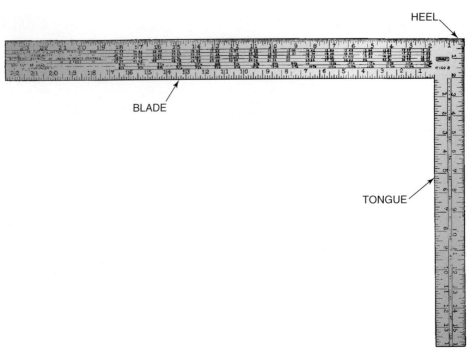

Figure 4-13: **Framing square**

Combination Squares

A combination square (Figure 4-14) is a most versatile tool. It has a movable handle that can lock in place on the 12" steel rule. It is used to square the end of a board, mark a 45° angle for mitering, and even make quick level checks with the built-in spirit level. It can also be used as a scribing tool to mark a constant distance along the length of a board.

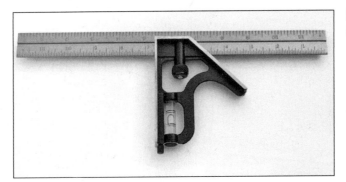

Figure 4-14: **Combination square**

Chalkline Reel

A chalkline is a string or line coated with colored chalk and is used to transfer a straight line to a working surface easily and accurately (see Figure 4-15). To use it, pull the line out and hold it tight between the two points of measurement; then snap it to leave a mark (shown in Figure 4-16). Some have a pointed case to double as a plumb bob.

Levels

A level (Figure 4-17) is used to make sure your work is true horizontal (level) or true vertical (plumb).

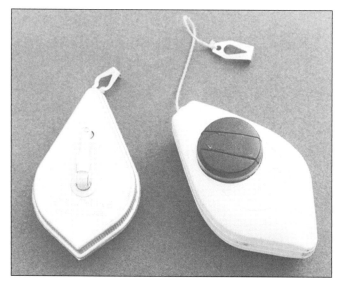
Figure 4-15: **Chalkline reel**

Figure 4-16: **Snapping a chalk line**

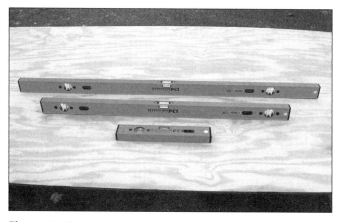

Figure 4-17: A spirit level has one or more transparent vials containing a liquid and a bubble

Plumb Bob

A plumb bob (Figure 4-18) is a heavy, balanced weight on a string, which you drop from a specific point to locate another point exactly below it, or to determine true vertical.

Boring and Cutting Tools

Boring and cutting tools are a category of tools used for making holes or cutting wood, metals, and plastic.

Drill Bits

To drill a satisfactory hole in any material, the correct type of drill bit must be used; it must be used correctly and be sharpened as appropriate. A set of high-speed steel twist drills and some masonry bits will probably be sufficient for the average facilities technician.

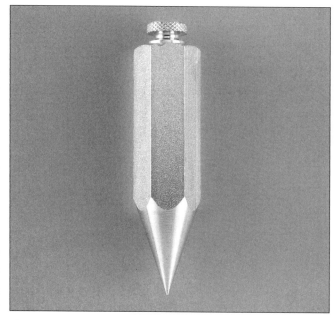
Figure 4-18: **Plumb bob**

Types of drill bits include:

- **Twisted bit**—the most common drilling tool used with either a hand or electric drill (see Figure 4-19). It can be used on timber, metal, plastic, and similar materials. Most twist bits are made from either:
  - **High speed steel** (HSS)—suitable for drilling most types of material. When drilling metal, the HSS stands up to the high temperatures.
  - **Carbon steel**—specially ground for drilling wood and should not be used for drilling metals. Carbon steel bits tend to be more brittle and less flexible than HSS bits.
- **Masonry bit**—designed for drilling into brick, block, stone, quarry tiles, or concrete. These bits are normally used in power drills (see Figure 4-19).
- **Spade bit**—used to drill fairly large, flat-bottomed holes. A spade bit is inexpensive and suitable for general work; however, they do not have good chip clearing ability and tend to split thin material (see Figure 4-19).
- **Hole saw**—used for cutting large, fixed, diameter holes in wood or plastic (see Figure 4-20).

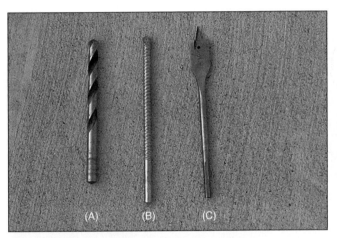

Figure 4-19: Common drill bits: (A) twist drill bit, (B) masonry bit, and (C) spade bit

Figure 4-20: Hole saw

Drilling tips:

- Always wear eye protection.
- Don't apply too much pressure on small drill bits.
- Ease up on pressure when drill breaks through material.
- The larger the drill bit, the slower the speed of the drill.
- Use a vise or clamp to hold the material to prevent it from spinning if the drill bit catches.

Saws

There are various types of saw, such as the rip, crosscut, and the panel saw. They all look basically the same and their primary purpose is the cutting of timber from boards, and sometimes making larger joints.

Types of saws include:

- **Crosscut saw**—available in many sizes and configurations. The crosscut saw is a good general purpose saw typically used for cutting wood across its grain (see Figure 4-21). The kerf, or the actual cut made by the saw, is as wide as the set of teeth in the saw blade.

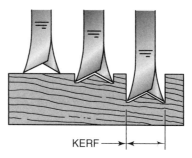

Figure 4-21: **Crosscut saw**

- **Ripsaw**—used to cut with the grain of the wood (see Figure 4-22). Ripsaw teeth are filed straight across the blade, and so each tooth is shaped like a little chisel.

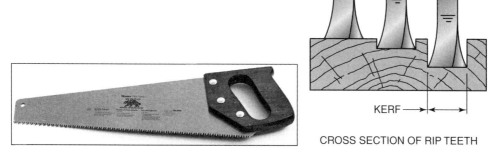

Figure 4-22: **Ripsaw**

- **Hacksaw**—basic hand saw used mostly for cutting metal (see Figure 4-23). Some have pistol grips, which make the hacksaw easy to grip. Hacksaws cut in straight lines.

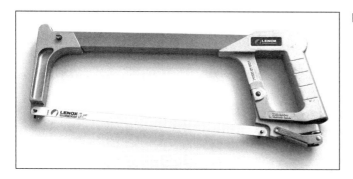

Figure 4-23: **Hacksaw**

- **Wallboard saw**—used for cutting electrical outlet holes and other small rough sawing where a powered saber saw will not fit (see Figure 4-24). A self-starting keyhole saw (a type of wallboard saw) is very handy and comfortable to use.
- **Coping saw**—used to cut the profile of one piece of molding on the end of molding (see Figure 4-25).

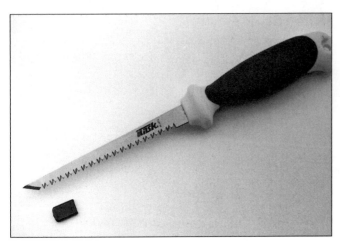

Figure 4-24: Wallboard saw

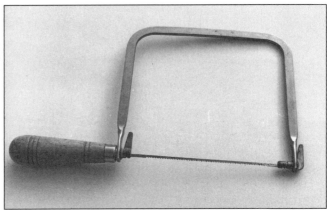

Figure 4-25: Coping saw

Hammers

Hammers are used for driving nails, fitting parts, and breaking up objects.

Types of hammers include:

- Sledge hammer—mainly used on outdoor projects. These hammers are designed to deliver heavy force (see Figure 4-26).
- Mason's hammer—used for working on brick, concrete, or mortar (see Figure 4-27). This hammer is often used for cutting and setting brick.

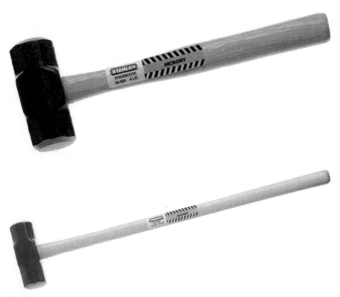

Figure 4-26: Sledge hammers

Figure 4-27: Mason's hammer

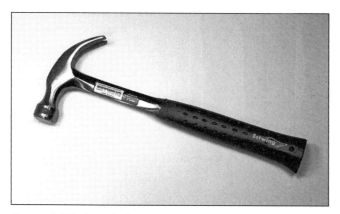 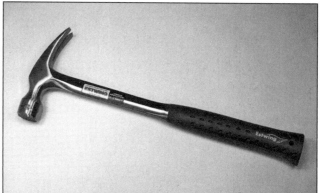

Figure 4-28: **Curved-claw hammer and framing hammer**

- Claw hammer—a general use hammer (see Figure 4-28).

  Tips for using a hammer:

- If working with hard wood, drill a pilot hole in the material to prevent splitting.
- To begin hammering, grip the hammer firmly in the middle of the handle and shake hands with your hammer!
- Don't hold the hammer too tightly.
- Hold the nail between the thumb and forefinger of the other hand and place the nail where it is to be driven.
- Using the center of the hammer face, drive the nail in with firm, smooth blows.
- The striking face should always be parallel with the surface that's being hit.
- Avoid sideways or glancing blows.
- Take care not to mark the work surface.

Screwdrivers

Screwdrivers (Figure 4-29) are used to insert and tighten, or to loosen and remove, screws. To select the proper screwdriver for a particular job, match the appropriate screw driver and size to the screw head (see Figure 4-30).

Types of screwdrivers include:

- slotted
- Phillips
- Torx
- square

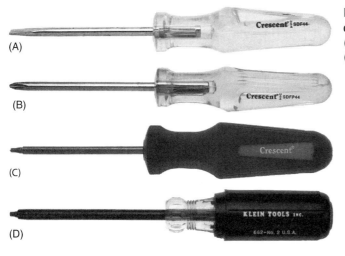

Figure 4-29: **Screwdrivers: (A) sloted, (B) Phillips, (C) torx, (D) Square or Robertson**

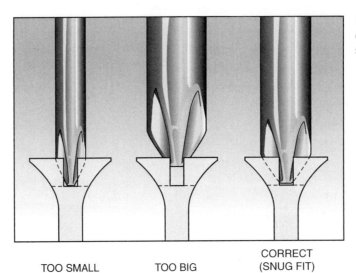

Figure 4-30: The screwdriver must be the right size to fit the fastener

TOO SMALL TOO BIG CORRECT (SNUG FIT)

Figure 4-31: Pliers: (A) common slip-joint, (B) needle nose, (C) channel-lock, (D) vise grip, (E) side-cutting, (F) electrician's

Pliers

Pliers are used primarily for gripping objects that utilize leverage. Pliers are designed for numerous purposes and require different jaw configurations to grip, turn, pull, or crimp a variety of things (see Figure 4-31).

Wrenches

Wrenches are used to turn bolts, nuts, or other hard-to-turn items.

Some common types of wrenches include:

- Socket wrench—has a ratchet handle, which allows the user to move the handle back and forth without having to take the socket off the nut and reposition it. This wrench uses separate, removable sockets to fit many different sizes of nuts (see Figure 4-32).
- Open-end wrench—usually has different sizes at each end (see Figure 4-33).

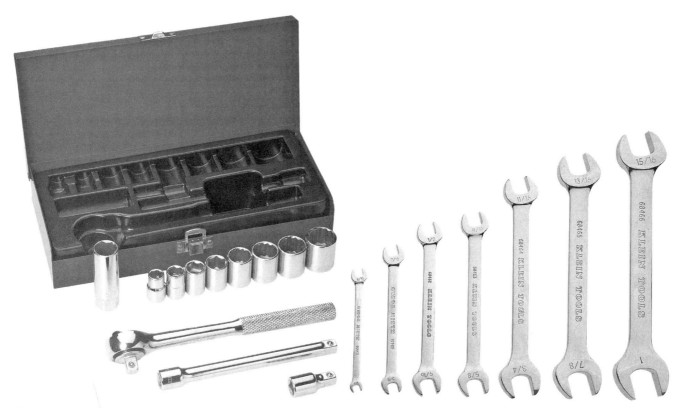

Figure 4-32: Socket wrench set

Figure 4-33: Open-ended wrenches

- Box-end wrench—usually has different sizes at each end that form a complete circle and either 6 or 12 points (see Figure 4-34).
- Adjustable wrench—has jaws that can be adjusted by turning the adjusting screw (see Figure 4-35).

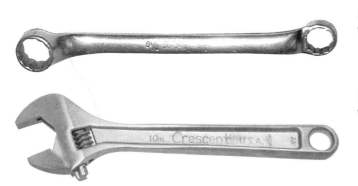

Figure 4-34: Box-end wrench

Figure 4-35: Adjustable wrench

Portable Power Tools

The term **power tool** is used to describe a tool that contains a motor. Power tools are used for some specific purposes or operations that one cannot perform manually. Portable tools are easy to move from one location to another.

Figure 4-36: Circular saw

Power Saws

Power saws are more commonly used in construction because they provide a means of cutting a material quickly and more efficiently than a hand saw.

Types of power saws include:

- Circular saw—the most basic portable power saw (see Figure 4-36). Most of these saws can be equipped with a rip guide to maintain a uniform width of cut on long passes.
- Reciprocating saw—similar to a non-powered hand saw except that the blade moves back and forth under power (see Figure 4-37). It is good for ripping and cross-cutting, but lacks the control achieved from the platform design of the saber saw.
- Saber saw—has a small, thin blade that cuts with an up-and-down motion, making it ideal for irregularities and scroll work, as well as for ripping and cross-cutting (see Figure 4-38).

Tips for using saws:

1. Read the owner's manual carefully before operating your power saw.
2. Observe all of the safety precautions discussed in the owner's manual.
3. Use the safety devices that come with your saw.
4. Keep your hands away from the blade.
5. Stand to one side of the blade while it is in motion in case wood is kicked back.
6. Unplug the saw before changing blades or making adjustments.
7. Be sure that the blades you use are sharp and clean to avoid binding and burning.

Figure 4-37: Reciprocating saw

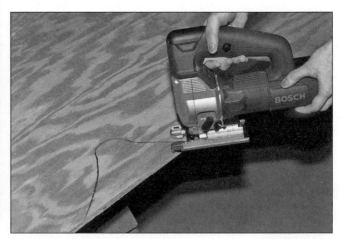

Figure 4-38: Saber saw

Drills and Drivers

The most basic of the electric drills, the rotary drill is used mainly for boring holes in a variety of materials. These can be either corded or cordless. Drivers are similar to the rotary drill but with a greater torque, allowing the user to drive and remove screws as well as drill through materials at a more rapid rate (see Figure 4-39).

Planes

Planes are use for planing rough boards and are ideal for trimming doors and frames for a perfect fit (see Figure 4-40).

Tips on how to plane:

Figure 4-39: **Portable power drills**

- As with all power tools, follow the manufacturers' safety instructions carefully.
- Place the front of the plane on the end of the wood, but make sure the cutter block is not touching the timber. Gently press down on the front handle and turn on the planer. Move the tool at an even rate along the wood. When you reach the end, transfer pressure to the rear handle and glide off the wood to avoid taking a deep gouge out of the last few millimeters of the work. Choose a portable planer that can be inverted in its own accessory stand if you want to plane several small pieces.
- To even out a wide surface, set the planer to its finest cut and plane diagonally to the grain, in overlapping strokes. You will still need to use a hand plane to get rid of slight machining marks.
- Make several light passes rather than taking off a lot of timber in one go.

Figure 4-40: **Portable electric plane**

Routers

Routers can be used to make raised-panel doors, round the edge of a coat rack, or create your own baseboard molding. The router can also trim plastic laminate.

There are two types of router: the plunge router and the fixed (or standard) router. Both types offer the same results, although each type is better for particular jobs.

1. Fixed router—used to make many different cuts including grooves, dadoes, rabbets, and dovetails. It is also used to shape edges and make cutouts (see Figure 4-41).
2. Plunge router—are used for a variety of different applications in which a fixed router cannot be used. A typical application for a plunge router in construction is to trim the edges of plastic laminates (see Figure 4-42). There are special router blades for finishing plastic laminates. The most popular are a flush cut blade and a beveled blade.

Figure 4-41: **Portable electric router**

Figure 4-42: **Laminate router**

Figure 4-42a: **A guide attached to the base of the router rides along the edge of the stock and controls the sideways motion of the router**

A tip for using the router: The edge guide has a straight face that can be adjusted at different distances from the bit. You adjust it so that it runs against a straight edge of the work piece, and the edge guide keeps the router going in a straight line (see Figure 4-42a).

Four basic types of router bits (Figure 4-43) are:

- Grooving bits—make a groove in the piece of wood. This type of bit is commonly used for street address signs for homes. Different types of grooving bits include the V-groove, the round-nose, and the straight bit.
- Joinery bits—make several different types of joints. This type of router bit includes the finger joint, the drawer lock, the rile and stile, and dovetail bits.

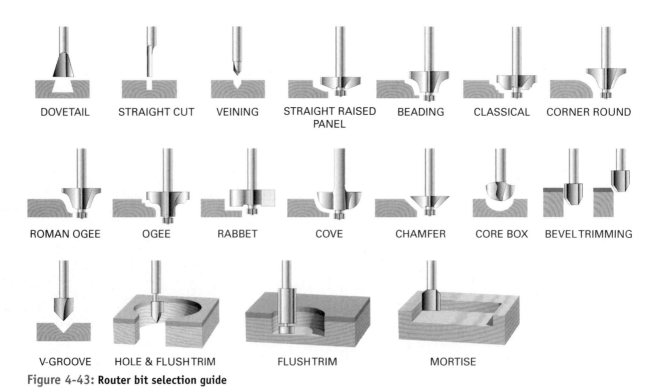

Figure 4-43: **Router bit selection guide**

- Edge bits—used to create different-shaped edges in woodwork. Examples of these types of bits include the beading, flush, and round-over bits.
- Specialized bits—do not fit into one of the above categories and have more specialized purposes, including the key hole, raised panel, and T-slot bits.

Sanders

Sanders can remove large amounts of stock or finishing material quickly and put a glass-smooth surface on your projects.

Types of sanders include:

- **Detail sanders**—small handheld sanders designed for sanding around odd shapes and small nooks in woodwork (see Figure 4-44).
- **Random orbit sanders**—random motion lets the operator move the sander in any direction without scarring the work surface. Random orbit sanders give you a superfine finish and leave a minimum amount of marks (see Figure 4-45).
- **Belt sanders**—used for removing paint or varnish from large areas, smoothing out rough wood, and preparing surfaces for finishing and thinning out thick wood (see Figure 4-46).
- **Disc sanders**—bench-mounted tools with a circular pad that accepts specially made sanding sheets. Most disc sanders also have a belt mounted vertically or horizontally on their frame (see Figure 4-47).

Figure 4-44: Detail sander (©2006 JupiterImages Corporation)

Figure 4-45: Random orbit sander (©2006 JupiterImages Corporation)

Figure 4-46: Belt sander

Figure 4-47: Disc sander

Figure 4-48: **Spindle sander**

- **Spindle sanders**—bench-mounted tools with a cylindrical spindle located in the center of a large worktable. The spindle holds special sanding tubes of various grit sandpapers. Some spindle sanders have an oscillating feature that raises and lowers the spindle as it rotates. The oscillating feature increases the rate at which the sander removes stock. Spindle sanders are good for edge sanding, especially around curves and circles (see Figure 4-48).

Sandpaper is a sheet abrasive composed of particles of flint, garnet, emery, aluminum oxide, or silicon carbide. These particles are mounted on paper or cloth in "open coat" or "closed coat" density (see Figure 4-49). Types of sandpaper include:

Grit	Common Name	Uses
40–60	Coarse	Heavy sanding and stripping, roughing up the surface.
80–120	Medium	Smoothing of the surface, removing smaller imperfections and marks.
150–180	Fine	Final sanding pass before finishing the wood.
220–240	Very Fine	Sanding between coats of stain or sealer.
280–320	Extra Fine	Removing dust spots or marks between finish coats.
360–600	Super Fine	Fine sanding of the finish to remove some luster or surface blemishes and scratches.

Figure 4-49: **Sample sandpaper grit table. The grit range may vary depending on manufacturers.**

- **Flint**—the least expensive sandpaper sold, this is a grey material that wears down quickly.
- **Garnet**—a much harder grit than flint and more suitable for woodworking, costs slightly more than flint paper.
- **Emery**—has a distinctive black color and is generally used on metal.
- **Aluminum oxide**—has a reddish-colored, very sharp grit and is used on either wood or metal.
- **Silicon carbide**—this bluish-black material is the hardest of all, and is commonly used for finishing metal or glass.

Sandpaper grit is identified by numbers from 1500 to 12; the smaller the number, the coarser the grit.

Stationary Tools

Stationary tools tend to be larger, more powerful, and more complex than hand tools.

- **Miter**—replaces the miter block used with a hand saw, and the various miter saw mechanisms that attempt to support a manual saw in a framework. It excels at any cut in which a precise angle is required. Typical applications include cutting

studwork, cutting miters for picture frames, dados, skirting, architraves, and coves. Also used in all sorts of general-purpose carpentry and joinery (see Figure 4-50).
- Chop—a lightweight circular saw mounted on a spring-loaded pivoting arm and supported by a metal base. Chop saws are considered the best saw to get very exact, square cuts (see Figure 4-51).
- Band saw—capable of performing a whole range of cuts, such as ripping, cross cutting, beveled cuts, and curves. The band saw is also capable of re-sawing, cutting a thick board into several thinner boards (see Figure 4-52).

Tips for using the band saw:

- Always stand to the left of the band saw. In the event of a broken blade, the blade will fly off to the right. If the blade breaks, shut off the power and stay away from the saw until it stops.

Figure 4-50: Miter saw

Figure 4-51: Chop saw

Figure 4-52: Band saw

- Use gloves when uncoiling, removing, and installing the band saw blade.
- Keep your hands and fingers away from exposed parts of the blade.
- Follow the manufacturer's guidelines for adjustment of the sliding bar or post. If the guide is too high, the blade will not have the proper support.
- Avoid backing out of the cut. This could push the blade off the wheels.
- Table saws—the superstar of stationary woodworking power tools. It rips, miters, bevels long edges, crosscuts, and mills nearly every imaginable woodworking joint. The table saw size is determined by the diameter of the largest blade that can be used on the saw. A 10" table saw can use up to a 10" saw blade (see Figure 4-53). Tips for using the table saw include:
  - Use the saw guard at all times. No operation should be done with the guards removed.
  - Never reach over the saw blade to remove scraps, or to provide support to the work piece.

Figure 4-53: A table saw

  - Always stand to the side of the saw, and never directly in line with the blade. If the saw catches the material you are working on, the saw will throw it in line with the blade.
  - When cutting, NEVER PULL the work piece through the saw. Start and finish the cut from the front of the saw.
  - When crosscutting, hold the work piece firmly against the miter gauge. Make sure that the miter gauge works freely in the slot and that it will clear both sides of the blade when tilted. Note that on some saws the miter gauge can be used only on one side when the blade is tilted.
  - Use a push stick according to the manufacturer's guidelines.
- Drill press—more accurate than any portable drill, a drill press uses a drilling head positioned above an adjustable workbench, both being fixed to a sturdy base. Most models include a clamp and a guide, allowing the user greater control when drilling (see Figure 4-54).

Tips for using the drill press include:

- Always secure the material being drilled.
- Use bits designed only for the drill press.

Figure 4-54: Drill press (©2006 JupiterImages Corporation)

- Never try to stop the machine by taking hold of the chuck after the power is off.
- Check to make sure the chuck is secured before turning the drill press on.

Safety tips for using stationary tools include:

- Safety devices and guards must always be in place. These devices were designed by the manufacturer to be used with the tool.
- Always keep blades and cutting edges sharp.
- Perform maintenance, accessory changes, and adjustments only when the tool is off and unplugged.
- Don't wear loose fitting clothing. High-powered stationary tools can catch clothing and draw the operator's body into the tool.
- When using any type of stationary saw, never use gloves. They can get caught in the saw.
- Never put your fingers and hands in front of saw blades and other cutting tools.
- Read the owner's manual and safety precautions before using the tools.
- Know the location of emergency on and off switches and the tools power switch.
- Make sure that blades, bits, and accessories are properly mounted. In addition, make sure all locking handles and clamps are tight before using a tool.

Tools, whether hand tools or power tools, can be dangerous if proper safety precautions are not followed. When working on projects that require the use of these tools, kindly ask the client to stay a safe distance from you while you are working and use all of the possible safety precautions.

Review Questions

1. What are fasteners commonly used for?

2. Define the following:
   Common nail
   Box nail
   Finishing nail

3. When is a screw used?

4. When are adhesives used?

5. What is a solvent?

6. List three types of saws commonly used today in facility maintenance.

7. How does a claw hammer differ from a mason's hammer?

8. How does an open-end wrench differ from a box-end wrench?

9. List two types of routers commonly used today.

10. List the four basic types of router bits commonly used today.

Name: _____

Date: _____

Tools and Equipment Job Sheet 1

Tools and Equipment

Fasteners

Upon completion of this job sheet, you should be able to demonstrate your ability to select the correct fastener for the appropriate repair project.

1. Take an inventory and list the different types of fasteners used at your facilities. Write a brief description of where they would be used.

   _____
   _____

2. Are the correct fasteners used for the appropriate jobs at your facilities?
   a. If no, what, if anything, should be done?

   _____
   _____

3. What are the different types of screws available and what determines which ones are used?

   _____
   _____

4. Write a brief explanation of what determines when a nail should be used versus when a screw should be used.

   _____
   _____

Instructor's Response:

_____
_____
_____

Name: _____

Date: _____

Tools and Equipment Job Sheet

2

Tools and Equipment

Adhesives

Upon completion of this job sheet, you should be able to demonstrate your ability to select the correct adhesive for the appropriate repair project.

1. Write a brief explanation of what determines when to use an adhesive.
   _____
   _____

2. Take an inventory of the types of adhesives used at your facility. Briefly describe where the different adhesives should be used.
   _____
   _____

Instructor's Response:
_____
_____
_____
_____

Name: _____

Date: _____

Tools and Equipment Job Sheet

3

Tools and Equipment

Hand Tools

Upon completion of this job sheet, you should be able to demonstrate your ability to safely use hand tools.

1. Compare four different brands of tape measures.
   a. How are the symbols different from one tape measure to another and what do the symbols mean?

   _____
   _____

   b. What are the increments of the scales?

   _____
   _____

2. Check your combination square for accuracy.
   a. Tape a sheet of paper to a board that has a perfectly straight edge.
   b. Hold the square against the edge and draw a line along the outer edge of the blade.
   c. Flip the square over so the opposite side of the blade faces up, align the square on the edge of the stock, and draw a second line about $1/32$" from the first.
   d. If the square is accurate, the lines will be parallel.

3. Investigate the different types of drill bits available at your facility. Write a brief description of the differences and the materials they are used on.

   _____
   _____

4. Investigate the different types of hand saws available at your local home improvement center and write a brief description of the physical differences of the saws and how each one should be used.

   _____
   _____

5. Compare and contrast the uses of the portable power tools discussed in this course with their comparable hand tools and/or stationary tools. Discuss how you decide which tool is most appropriate for the project.

Instructor's Response:

_____
_____
_____
_____

Chapter 5 Practical Electrical Theory

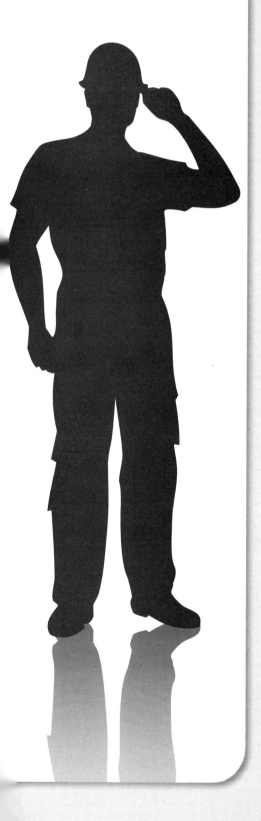

OBJECTIVES

By the end of this chapter, you will be able to:

Knowledge-Based

- Understand the basic principle of basic electricity.
- Describe the difference between AC and DC currents.
- Understand properties of common electrical wires used by facilities maintenance technicians and understand and correctly measure wire size and load carry capacity.
- Understand operation and functions of emergency circuits.
- Describe different types of emergency back-up electrical power systems.

Skill-Based

- Calculate electrical load by using Ohm's law.

Introduction

Electricity is the driving force that provides most of the power for the industrialized world. It is used to light homes, cook meals, heat and cool buildings, and drive motors, and serves as the ignition for most automobiles. This module provides the technician with a basic understanding of practical electrical theory.

Basic Electrical Theory

Although the practical use of electricity has become common only within the last hundred years, it has been known as a force for much longer. The Greeks were the first to discover electricity, about 2,500 years ago. They noticed that when amber was rubbed with other materials, it became charged

Figure 5-1: **Three stages of water**

with an unknown force that had the power to attract objects such as dried leaves, feathers, bits of cloth, or other lightweight materials. The Greeks called amber *elektron*. The word "electric" was derived from it, meaning "to be like amber" or to have the ability to attract other objects.

To understand electricity, it is necessary to start with the study of atoms. The **atom** is the basic building block of the universe. All matter is made by combining atoms into groups. **Matter** is any substance that has mass and occupies space. It can exist in any of three states: solid, liquid, or gas. Water, for example, can exist as a solid in the form of ice, as a liquid, or as a gas in the form of steam (see Figure 5-1).

An **element** is a substance that cannot be chemically divided into a simpler substance (see Figure 5-2).

An atom is the smallest part of an element. It is composed of three principal parts: protons, neutrons, and electrons. The proton is the positively-charged particle that combines with the neutron to form the nucleus of the atom. Because the neutron has no charge, the nucleus will have a net positive charge. However, the electron is a negatively-charged particle that orbits around the nucleus of the atom. Typically an atom will contain the same number of electrons and protons, thus producing a neutrally charged atom (see Figure 5-3).

To understand atoms, it is necessary first to understand two basic laws of physics. One of these is the **law of charges**, which states that opposite charges attract and like charges repel (see Figure 5-4).

If two objects with unlike charges come close to each other, the lines of force attract (Figure 5-5). Likewise, if two objects with like charges come close to each other, the lines of force repel (Figure 5-6). For example, consider what happens if you try to connect two north poles of a magnet. It is this principle that helps the electrons to maintain their orbit around the nucleus.

The second law that is important to understanding the behavior of an atom is the **law of centrifugal force**, which states that a spinning object will pull away from its center point and that the faster it spins, the greater the centrifugal force becomes (see Figure 5-7). If you tie an object to a string and spin it around, it will try to pull away from you. The faster the object spins, the greater the force that tries to pull the object away.

ATOMIC NUMBER	NAME	VALENCE ELECTRONS	SYMBOL	ATOMIC NUMBER	NAME	VALENCE ELECTRONS	SYMBOL	ATOMIC NUMBER	NAME	VALENCE ELECTRONS	SYMBOL
1	HYDROGEN	1	H	37	RUBIDIUM	1	Rb	73	TANTALUM	2	Ta
2	HELIUM	2	He	38	STRONTIUM	2	Sr	74	TUNGSTEN	2	W
3	LITHIUM	1	Li	39	YTTRIUM	2	Y	75	RHENIUM	2	Re
4	BERYLLIUM	2	Be	40	ZIRCONIUM	2	Zr	76	OSMIUM	2	Os
5	BORON	3	B	41	NIOBIUM	1	Nb	77	IRIDIUM	2	Ir
6	CARBON	4	C	42	MOLYBDENUM	1	Mo	78	PLATINUM	1	Pt
7	NITROGEN	5	N	43	TECHNETIUM	2	Tc	79	GOLD	1	Au
8	OXYGEN	6	O	44	RUTHENIUM	1	Ru	80	MERCURY	2	Hg
9	FLUORINE	7	F	45	RHODIUM	1	Rh	81	THALLIUM	3	Tl
10	NEON	8	Ne	46	PALLADIUM	–	Pd	82	LEAD	4	Pb
11	SODIUM	1	Na	47	SILVER	1	Ag	83	BISMUTH	5	Bi
12	MAGNESIUM	2	Ma	48	CADMIUM	2	Cd	84	POLONIUM	6	Po
13	ALUMINUM	3	Al	49	INDIUM	3	In	85	ASTATINE	7	At
14	SILICON	4	Si	50	TIN	4	Sn	86	RADON	8	Rd
15	PHOSPHORUS	5	P	51	ANTIMONY	5	Sb	87	FRANCIUM	1	Fr
16	SULFUR	6	S	52	TELLURIUM	6	Te	88	RADIUM	2	Ra
17	CHLORINE	7	Cl	53	IODINE	7	I	89	ACTINIUM	2	Ac
18	ARGON	8	A	54	XENON	8	Xe	90	THORIUM	2	Th
19	POTASSIUM	1	K	55	CESIUM	1	Cs	91	PROTACTINIUM	2	Pa
20	CALCIUM	2	Ca	56	BARIUM	2	Ba	92	URANIUM	2	U
21	SCANDIUM	2	Sc	57	LANTHANUM	2	La				
22	TITANIUM	2	Ti	58	CERIUM	2	Ce	ARTIFICAL ELEMENTS			
23	VANADIUM	2	V	59	PRASEODYMIUM	2	Pr				
24	CHROMIUM	1	Cr	60	NEODYMIUM	2	Nd	93	NEPTUNIUM	2	Np
25	MANGANESE	2	Mn	61	PROMETHIUM	2	Pm	94	PLUTONIUM	2	Pu
26	IRON	2	Fe	62	SAMARIUM	2	Sm	95	AMERICIUM	2	Am
27	COBALT	2	Co	63	EUROPIUM	2	Eu	96	CURIUM	2	Cm
28	NICKEL	2	Ni	64	GADOLINIUM	2	Gd	97	BERKELIUM	2	Bk
29	COPPER	1	Cu	65	TERBIUM	2	Tb	98	CALIFORNIUM	2	Cf
30	ZINC	2	Zn	66	DYSPROSIUM	2	Dy	99	EINSTEINIUM	2	E
31	GALLIUM	3	Ga	67	HOLMIUM	2	Ho	100	FERMIUM	2	Fm
32	GERMANIUM	4	Ge	68	ERBIUM	2	Er	101	MENDELEVIUM	2	Mv
33	ARSENIC	5	As	69	THULIUM	2	Tm	102	NOBELIUM	2	No
34	SELENIUM	6	Se	70	YTTERBIUM	2	Yb	103	LAWRENCIUM	2	Lw
35	BROMINE	7	Br	71	LUTETIUM	2	Lu				
36	KRYPTON	8	Kr	72	HAFNIUM	2	Hf				

Figure 5-2: **Table lists both natural and artificial elements**

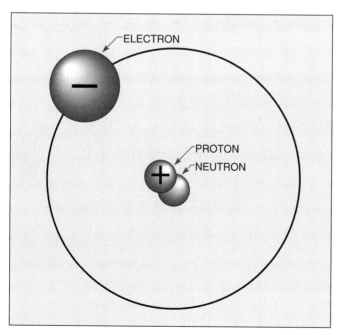

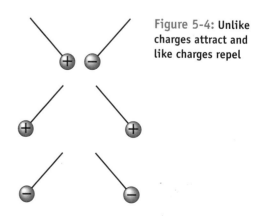

Figure 5-3: **Parts of the atom**

Figure 5-4: **Unlike charges attract and like charges repel**

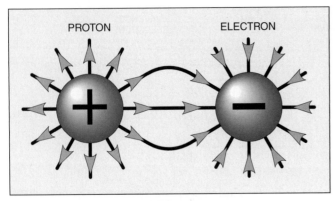

Figure 5-5: Unlike charges attract each other

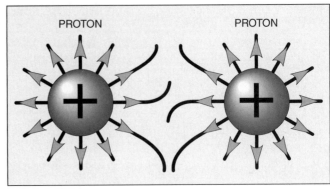

Figure 5-6: Like charges repel each other

Figure 5-7: Centrifugal force causes an object to pull away from its axis point

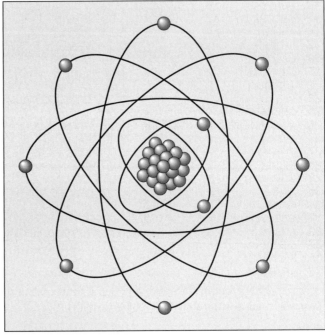

Figure 5-8: Electron orbits

Although atoms are often drawn flat, electrons orbit the nucleus in a spherical fashion (Figure 5-8). Electrons travel at such a high rate of speed that they form a shell around the nucleus. For this reason, electron orbits are often referred to as shells.

The number of electrons that can be contained in any one orbit, or shell, is found by using the formula ($2N^2$). The letter N represents the number of the orbit, or shell (see Figure 5-9).

The outer shell of an atom is known as the **valence shell**. Any electrons located in the outer shell of an atom are known as valence electrons (Figure 5-10). The valence shell of an atom cannot hold more than eight electrons. The valence electrons are of primary concern in the study of electricity, because they explain much of electrical theory. A conductor, for instance, is generally made from a material that contains one or two valence electrons. Atoms with one or two valence electrons are unstable and can be made to give up these electrons with little effort. Conductors are materials that

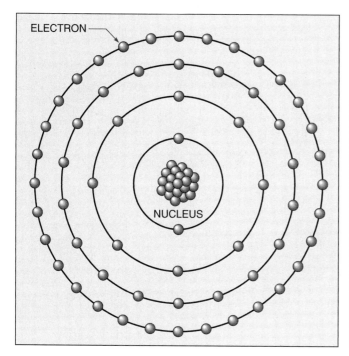

Figure 5-9: Electrons orbit the nucleus in a circular fashion

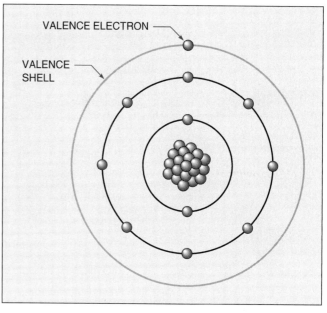

Figure 5-10: The electrons located in the outer orbit of an atom are valence electrons

permit electrons to flow through them easily. Examples of good conductors are gold, silver, copper, and aluminum.

Materials containing seven or eight valence electrons are known as insulators. Insulators are materials that resist the flow of electricity. Some good examples of insulator materials are rubber, plastic, glass, and wood.

A **coulomb** is a quantity measurement for electrons. One coulomb contains 6.25×10^{18}, or 6,250,000,000,000,000,000, electrons. Coulomb's law of electrostatic charges states that the force of electrostatic attraction or repulsion is directly proportional to the product of the two charges and inversely proportional to the square of the distance between them.

The **amp**, or **ampere**, is named for André Ampére, a scientist who lived from the late 1700s to the early 1800s. The amp (A) is defined as one coulomb per second. One amp of current flows through a wire when one coulomb flows past a point in one second (Figure 5-11).

The ampere is a measurement of the amount of electricity that is flowing through a circuit.

There are actually two theories concerning current flow: the theory known as the electron theory and the conventional current flow theory. The **electron theory** states that since electrons are negative particles, current flows from the most negative point in the circuit to the most positive. The **conventional current flow theory** is older

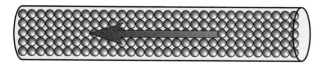

Figure 5-11: One ampere equals one coulomb per second

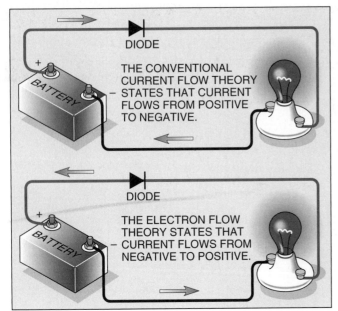

Figure 5-12: Conventional current flow theory and electron flow theory

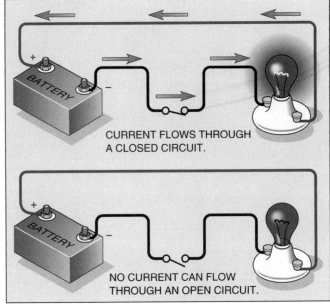

Figure 5-13: Current flows only through a closed circuit

than the electron theory and states that current flows from the most positive point to the most negative (see Figure 5-12).

A complete path must exist before current can flow through a circuit. A complete circuit is often referred to as a closed circuit because the power source, conductors, and load form a closed loop (see Figure 5-13). A short circuit, which has very little or no resistance, generally occurs when the conductors leading from and back to the power source become connected together (see Figure 5-14). Another type of circuit, one that is often confused with a short circuit, is a grounded circuit (see Figure 5-15). Grounded circuits can also cause an excessive amount of current flow. They occur when a path other than the one intended is established to ground. Many circuits contain an extra conductor called the grounding conductor. The grounding conductor helps prevent a shock hazard in the event the ungrounded, or hot, conductor comes in contact with the case or frame of the appliance.

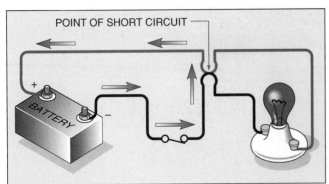

Figure 5-14: A short circuit bypasses the load and permits too much to flow

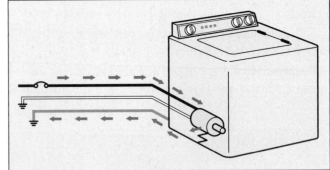

Figure 5-15: The grounding conductor provides a low-resistance path to ground

Voltage is defined as electromotive force, or EMF. It is the force that pushes the electrons through a wire and is often referred to as electrical pressure. The volt is the amount of potential necessary to cause one coulomb to produce one joule of work. An ohm is the unit of resistance to current flow. The symbol used to represent an ohm, or resistance, is the Greek letter omega (Ω). All electrical loads, such as heating elements, lamps, motors, transformers, and so on, are measured in ohms. **Wattage** is a measure of the amount of power that is being used in a circuit. Wattage is proportional to the amount of voltage and the amount of current flow. Watts is a measure of the amount of electrical energy converted into some other form.

Calculate Electrical Load by Using Ohm's Law

Ohm's law deals with the relationship between voltage and current and a material's ability to conduct electricity. Typically, this relationship is written as Voltage = Current × Resistance. In its simplest form, Ohm's law states that it takes one volt to push one amp through one ohm.

In a DC circuit, the current is directly proportional to the voltage and inversely proportional to the resistance.

The resistance, voltage, and/or current of a device (load) can be calculated by rearranging Ohm's law (Figure 5-16). The three basic forms of Ohm's law formulas are:

$E = I \times R$

Voltage can be found if the current and resistance are known.

$I = E/R$

Current can be found if the voltage and resistance are known.

$R = E/I$

Resistance can be found if the voltage and current are known.

Where

E = EMF, or voltage
I = intensity of current, or amperage
R = resistance

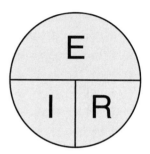

Figure 5-16: Chart for finding values of voltage, current, and resistance

The first formula states that the voltage can be found if the current and resistance are known. The second formula states that the current can be found if the voltage and resistance are known. The third formula states that if the voltage and current are known, the resistance can be found.

For example, determine the voltage for a load that draws 30 amps and has a resistance of 500 ohms (see Figures 5-17 and 5-18).

$E = I \times R$
$E = 30 \text{ Amps} \times 500 \text{ Ohms}$
$E = 15,000 \text{ Volts}$

Determine the current for a 120-volt load that has a resistance of 0.5 ohms.

$I = E/R$
$I = 120 \text{ Volts}/0.5 \text{ Ohms}$
$I = 60 \text{ Amps}$

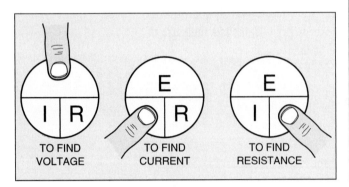

Figure 5-17: Using the Ohm's law chart

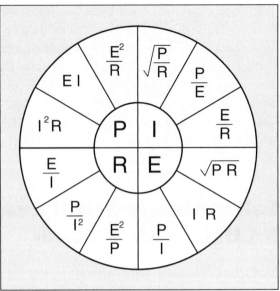

Figure 5-18: Formula chart for finding values or voltage, current, resistance, and power

Determine the resistance for a 120-volt load that draws 15 amps.

$R = E/I$
$R = 120$ Volts$/15$ Amps
$R = 8$ Ohms

As stated earlier, wattage is proportional to the amount of voltage and the amount of current flow. It can be calculated by using the formula: Watt = Voltage • Current. For example:

Determine the wattage of a 120-volt load that draws 15 amps.

$R = E/I$
$R = 120/15$ Amps
$R = 8$ Ohms

Watts $= E \times I$
Watts $= 120 \cdot 15$ Amps
Watts $= 1{,}800$ Watts

Difference between AC and DC Currents

Electricity flows in two ways: either as alternating current (AC) or as direct current (DC). The difference between AC and DC has to do with the direction in which the electrons flow. In DC, the electrons flow steadily in a single direction, or forward. In AC, electrons keep switching directions, or alternating.

AC electricity is the type of electricity commonly used in homes and businesses. AC electricity is commonly used to power lights, motors, and industrial, commercial, and residential processes and applications. However, when AC electricity is used to provide power to consumer electronics, a transformer is typically used to convert the AC electricity to DC electricity. AC electricity was proven to be better for supplying electricity than DC, primarily because the voltages can be transformed and can be transmitted over longer distances than DC electricity.

Correctly Measure Wire Size and Load Carry Capacity

The National Electrical Code (NEC) establishes important fundamentals that weave their way through the decision-making process for an electrical installation. The NEC defines **continuous load** as "a load where the maximum current is expected to continue for three hours or more." General lighting outlets and receptacle outlets in residences are not considered to be continuous loads.

When wiring a house, it is all but impossible to know which appliances, lighting, heating, and other loads will be turned on at the same time. Different families have different lifestyles. The rules for doing the calculations are found in Article 220 of the NEC. For lighting and receptacles, the computations are based on volt-amperes per square foot. For the small-appliance circuits in kitchens and dinning rooms, the basis is 1,500 volt-amperes per circuit. For large appliances such as dryers, electric ranges, ovens, cooktops, water heaters, air conditioners, heat pumps, and so on, which are not all used continuously or at the same time, there are demand factors to be used in the calculations. Following the requirements in the NEC, the various calculations roll together in steps that result in the proper sizing of branch-circuits, feeders, and service equipment.

Emergency Circuits

An emergency circuit is intended to supply illumination and power automatically to designated areas and equipment when the normal source of power fails. For the emergency circuit, a portable or temporary alternate source must be available whenever the emergency generator is out of service for major maintenance or repair.

Ground fault circuit interrupts (GFCI) are used to prevent people from being electrocuted. They work by sensing the amount of current flow on both the ungrounded (hot) and grounded (neutral) conductors supplying power to a device. A ground fault occurs when a path to ground, other than the intended path, is established (see Figure 5-19).

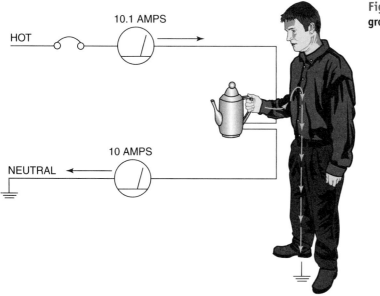

Figure 5-19: A ground fault

Ground-fault indication is required for emergency systems operating at more than 150 volts to ground and over-current devices rated at 1,000 amperes or more. Wiring for emergency circuits must be kept entirely independent of all other wiring unless required to be associated with normal source wiring.

Emergency Backup Electrical Power Systems

Emergency backup electrical power systems are used to provide continued access to electrical service during power outages. Systems currently available on the market make it possible to have continued access to electrical services during power outages. These systems can be based either on fossil-fuel-powered generators, on battery-based storage systems, propane or natural gas supply, or can be directly wired into the household circuit.

Permanent generators can be set up to power the whole building structure during an outage or just the essential loads, such as the furnace, security systems, and various appliances.

Review Questions

1. What are three stages of water?

2. What are the three principal parts of an atom?

3. State the law of charges.

4. What is a coulomb?

5. What is an amp?

6. What is electricity?

7. What is a watt?

Name: _____

Date: _____

Practical Electrical Theory Job Sheet 1

Practical Electrical Theory

Understanding Ohm's Law

Upon completion of this job sheet, you should have a basic understanding of Ohm's law.

1. Determine the voltage of a load drawing 20 amps and having a resistance of 25 ohms.

2. Determine the wattage of a device drawing 10 amps and having a resistance of 75 ohms.

Instructor's Response:

_____
_____
_____
_____

Chapter 6: Electrical Facilities Maintenance

OBJECTIVES

By the end of this chapter, you should be able to:

Knowledge-Based

- Understand and apply OSHA regulations that cover electrical installations.
- Describe the difference between AC and DC.
- Correctly identify single-phase and three-phase electrical systems.
- Correctly identify and select the boxes most commonly used in electrical installations.
- Correctly identify and select different types of electrical devices and fixtures.
- Describe the different types of emergency backup systems.

Skill-Based

- Follow systematic diagnostic and troubleshooting practices.
- Perform tests on smoke alarms, fire alarms, medical alert systems, and emergency exit lighting.
- Perform tests on GFCI receptacles.
- Repair and/or replace common electrical devices such as receptacles and switches.
- Repair and/or replace lighting fixtures and/or bulbs, and ballasts.

Introduction

Once the facility maintenance technician has a good understanding of basic electrical theory and electrical safety, the technician can attempt to troubleshoot, repair, and install basic electrical circuits and appliances. The previous chapter introduced the technician to basic electrical theory. This chapter introduces the concepts of safety and the basic procedures for troubleshooting and repairing basic electrical circuits and devices (switches, receptacles, and so on).

Safety, Tools, and Test Equipment

Safety

There are many safety considerations that must be adhered to while working on electrical systems, equipment, and devices. Many of the safety procedures are outlined in the OSHA standard (29 CFR Part 1926) and must be followed in order for work to be performed safely on electrical systems, equipment, and devices.

It is also important, in any installation, repair, or removal of electrical equipment or devices, that the proper wiring methods and practices in the National Electrical Code, better known as the NEC (NFPA 70), are used. The NEC is written to ensure the protection of persons and property from the hazards that arise from installation and/or repair of electrical systems or equipment. One must also follow all local codes and regulations that might exceed the minimum standards required by the NEC.

Prior to beginning any work on electrical systems, equipment, or devices, always consult with the local authority having jurisdiction (building/electrical inspector) on the local codes and regulations that might affect the proper and safe installation of electrical systems, equipment, and devices in your area.

The OSHA and NEC safety articles mandate that only qualified personnel should work on electrical systems, equipment, or devices. This is to protect not only the property owner, but also the person doing the installation. A qualified person is one who has the skills and knowledge related to the construction and operation of the electrical equipment and installations and has received safety training on the hazards involved.

Lock-out/tag-out procedures, as stated in the OSHA standard 1926.417 and the NFPA 70E standard, should be followed at all times to prevent electrical shock or even death by electrocution. There are many different forms of lockouts, Lock-outs can be placed on safety disconnects, switches, and breakers. You must use the lock-out that will prevent current from flowing through the circuit that you are working on. Some examples of the different lock-outs that are available can be seen in Figure 6-1.

The NFPA 70E mandates that qualified personnel should never work on any part of an energized electrical system that is over 50 volts AC (alternating current) without the proper arc flash/arc blast personnel protective clothing (PPE) and without using 1,000-volt-rated tools. This mandate is in effect for your protection only. Also, safety glasses should always be worn at all times while working on electrical systems, equipment, or devices.

Figure 6-1: **Examples of lock-out/tag-out devices**

Tools and Test Equipment

Always ensure that all hand tools and power tools are inspected before and after use. Look for any defects or damage that may cause injury while the hand tool or power tool is in use. If you notice any damage, the tool should not be used until it has been repaired by a qualified person (someone trained to work on the damaged tool). If a qualified person is not available to repair the damaged tool, the tool should be either discarded or put away until such a time when a qualified person can repair the tool. It is important to regularly maintain and clean all hand tools and power tools to ensure their proper operation.

Always use calibrated test equipment when testing electrical systems, equipment, or devices to ensure accurate measurements. Never use test equipment on any energized circuit above what it is rated for. This could cause the test equipment to explode, possibly leading to serious injury or even death.

Practical Electrical Theory

Before a technician can safely work with an electrical system, she should have a good understanding of the basic principles of electrical theory. A direct relationship exists between voltage, current, and resistance. Power is the product of voltage and current.

- **Voltage** is the amount of electrical pressure that is in a given circuit and is measured in *volts*.
- **Current** (or amperage) is the flow of electrons through a given circuit and is measured in *amps*.
- **Resistance** is the opposition to current flow in a given electrical circuit and is measured in *ohms*.
- **Power** is the electrical work that is being done in a given circuit and is measured in *wattage (watts)* for a purely resistive circuit or *volt-amps (VA)* for an inductive/capacitive circuit.

AC vs. DC

There are two types of electrical current used to power electrical circuits today: alternating current (AC) and direct current (DC). *Alternating current* is electrical current in which the magnitude and direction vary cyclically. *Direct current's* magnitude and direction does not modulate and therefore remains constant. Alternating current is primarily used in residential, commercial, and industrial applications as the primary source of power. Direct current is used primary in electronic, low-voltage applications, batteries, and so on.

Single-Phase AC vs. Three-Phase AC

Single-phase AC is the most commonly used electrical supply for single-family and multifamily dwellings. Single-phase consists of two ungrounded conductors ("hot wires") and one grounded conductor ("neutral wire"). When measuring voltage between the two hot wires of this type of system, you will read approximately 240 volts. If you read from any one hot wire to the neutral, you should read approximately 120 volts. Make sure that the meter that you are using to test voltage is rated appropriately.

Three-phase AC is most often used for commercial buildings, healthcare facilities, and industrial facilities. There are always three ungrounded conductors (hot wires) in this type of system. The three-phase system may or may not have a neutral wire. The common voltage levels for the three-phase system will be 208, 230, 240, 480, 575, or 600 when measured between the two hot wires. This is referred to as the line voltage. If a neutral is present, then voltage read from any one hot wire to the neutral wire will equal the line voltage divided by the square root of 3 (1.732).

Emergency Backup Systems

Emergency backup systems are designed to keep critical parts of an electrical system energized in the event of power loss. Hospitals and assisted living facilities have emergency backup systems in order that life support systems will not be interrupted during the loss of power. These types of emergency backup systems will typically use a diesel engine that drives a generator. The facility will switch over to the generator only in the event of power loss.

Another form of an emergency backup system is the uninterruptible power supply (UPS). (See Figure 6-2.) This system uses a battery or batteries to supply constant power to the circuits connected to it in the event of power loss. This system comes in a variety of sizes. A UPS can be small enough to fit on a desk or it can be so large that it may require its own room or even its own building on the facility grounds. The larger UPS systems require a lot of maintenance to maintain the reliability of the batteries. There are many safety issues to consider when working on or around these large battery backup systems and it is for this reason that only qualified personnel work on these systems.

Figure 6-2: **Uninterruptible Power Supply (UPS)**

Electrical Material, Equipment, Fixtures, and Devices

NM Cable

Using improper wire types and improper wire sizes can result in a fire! It is important to know what type of wire or cable is used for a given application.

The most common type of cable that is used in single-family and multi-family dwellings is NM cable (commonly referred to as Romex), NM cable is covered in article 334 of the NEC. This is where you would go to find the proper use and installation methods for NM cable. NM cable can be seen in Figure 6-3.

Figure 6-3: **NM cable**

NM cable comes in many sizes and conductor pairs. This means that an NM cable comes with two, three, or four current-carrying conductors and a ground. The ground that is present in NM cable is typically bare (without insulation); all of the current-carrying conductors will have an insulation covering that is identified by the following colors:

- Black
- Red
- Blue
- White

The blue insulated conductor is found in only four-conductor NM cable. Admittedly, a four-conductor NM cable with ground is very rarely used or seen, but it does exist. Most often, NM cable is sold as a two conductor with ground or a three conductor with ground.

Electrical Materials and Supplies

Boxes

There are a multitude of boxes being used today for electrical installations. Only the most common of these boxes will be discussed. One of the most common boxes in use is the single gang nail-up box. It is made of plastic or fiberglass and is typically used during new construction before the drywall is installed. A single-gang nail-up box can be seen in Figure 6-4.

This box will have only enough room for one yoked device, such as a switch, a receptacle, or a dimmer switch. A two-gang nail-up box will have enough room for two yoked devices. A three-gang nail-up box has enough room for three yoked devices.

The "cut-in" box (see Figure 6-5) is made of plastic or fiberglass and is used when a device must be installed after the drywall is already in place.

See page 90 for the procedure for installing old-work electrical boxes in a sheetrock wall or ceiling.

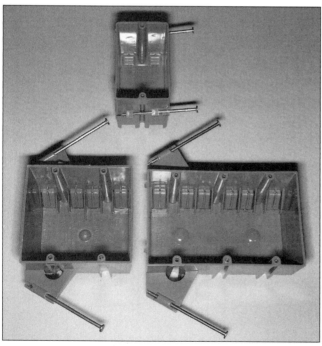

Figure 6-4: **Gang device boxes**

Figure 6-5: **Cut-in box**

Procedures

Installing Old-Work Electrical Boxes in a Sheetrock Wall or Ceiling

- Put on safety glasses and follow all applicable safety rules.

- Determine the location where you want to mount the box and make a mark. Make sure there are no studs or joists directly behind where you want to install the box.

- Turn the old-work box you are installing backward and place it at the mounting location so the center of the box is centered on the box location mark. Trace around the box with a pencil. Do not trace around the plaster ears.

- Using a keyhole saw, carefully cut out the outline of the box. There are two ways to get the cut started. One way is to use a drill and a flat blade bit (say a 1/2-inch size), drill out the corners, and start cutting in one of the corners. The other way that is used often is to simply put the tip of the keyhole saw at a good starting location and, with the heel of your hand, hit the keyhole saw handle with enough force to cause the blade to go through the sheetrock. It usually does not require much force to start the cut this way.

- **A** Assuming that a cable has been run to the box opening, secure the cable to the box and insert the box into the hole. Secure the box to the wall or ceiling surface with Madison hold-its or use a metal or nonmetallic box with built-in drywall grips.

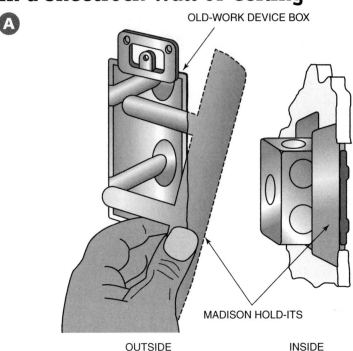

OUTSIDE WALL VIEW   INSIDE WALL VIEW

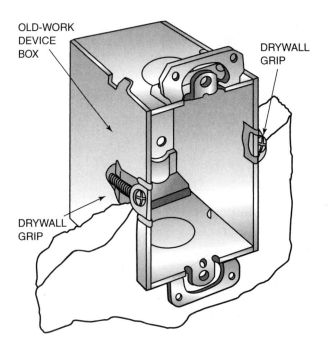

Installing a Cut-in Box

Round ceiling boxes (see Figure 6-6) come in a variety of diameters, depths, and shapes. There are four common types of ceiling boxes:

- pancake (the shallow box)
- 4" round nail-up box
- 4" round cut-in box
- fan-rated ceiling box

Ceiling fan boxes (see Figure 6-7) are used to support a lighting fixture in the ceiling. It is important to know that a ceiling fan should not be installed on a ceiling box that is not a fan-rated box. If a ceiling box is rated to support a fan, it will be clearly and legibly stamped into the interior of the box. If a fan is mounted to a box that is not rated for a ceiling fan, the fan could fall during operation and cause injury to someone underneath it.

Weatherproof boxes (see Figure 6-8) are used whenever a device (receptacle or switch) is installed in a wet location. This box is designed to keep the elements of the weather from affecting the electrical circuit and device that is contained within the

Figure 6-6: **Round ceiling box**

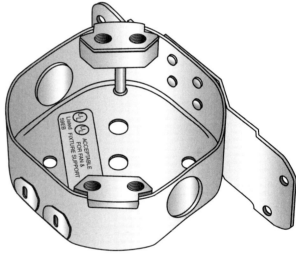

Figure 6-7: **Ceiling fan box**

Figure 6-8: **Weatherproof box and box with cover**

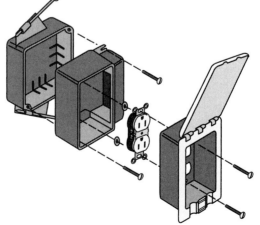

box. If this box is installed in the direct weather and it houses a receptacle, it is to have an "in-use" cover. An in-use cover looks like a giant clear bubble.

Electrical Devices and Fixtures

Electrical devices are switches and receptacles. There are many variations of each, so we will discuss only the most common of each.

Switches

Four of the most commonly used switches are the single-pole switch, the double-pole switch, the three-way switch, and the four-way switch. These switches are available in 15- and 20-amp ratings. Be sure to use the correct amp rating when installing a switch in a lighting branch circuit.

- Single-pole switch (see Figure 6-9): This switch is used when a light or fan is turned on or off from only one location.
- Double-pole switch (see Figure 6-10): This switch is used when two separate circuits must be controlled with one switch. They are used to control 240-volt loads, such as electric heat, motors, and electric clothes dryers.
- Three-way switch (see Figure 6-11): This switch is used when a light or fan can be turned on or off from two different locations.
- Four-way switch (see Figure 6-12): This switch is used when a light or fan can be turned on or off from three or more different locations. This switch must be used with two three-way switches.

Receptacles

There are many types of receptacles available. The most common receptacles are:

- 240-volt 30- or 50-amp single receptacle
- 240-volt 20-amp single receptacle
- 120-volt 15- or 20-amp single receptacle

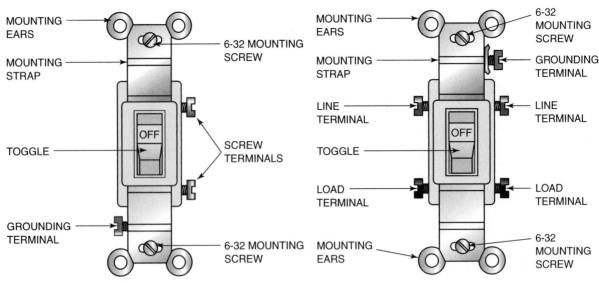

Figure 6-9: **Single-pole switch**

Figure 6-10: **Double-pole switch**

CHAPTER 6 Electrical Facilities Maintenance   93

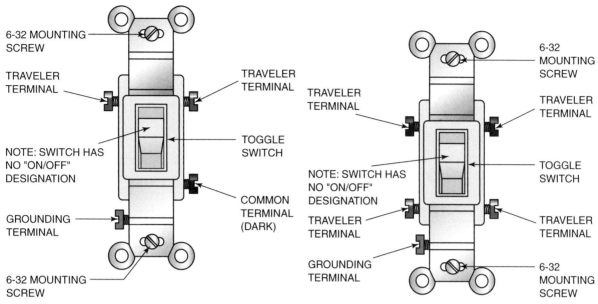

Figure 6-11: Three-way switch

Figure 6-12: Four-way switch

- 120-volt 15- or 20-amp duplex receptacle
- 120-volt 15- or 20-amp ground fault circuit interrupter (GFCI) receptacle

The *240-volt 30-amp single receptacle* is generally used as a clothes dryer receptacle. It has four wires that are connected to it; two ungrounded conductors (hot wires), one grounded conductor (neutral), and one grounding conductor (bare ground wire).

The *240-volt 50-amp single receptacle* is generally used as an electric range (stove) receptacle. It also has four wires that are connected to it; two ungrounded conductors (hot wires), one grounded conductor (neutral), and one grounding conductor (bare ground wire).

The *240-volt 20-amp single receptacle* is generally used as an air conditioner receptacle. As the previous receptacles, it also has four wires that are connected to it; two ungrounded conductors (hot wires), one grounded conductor (neutral), and one grounding conductor (bare ground wire). Notice the slots on the front of the receptacle.

The *120-volt 15/20-amp single receptacle* (see Figure 6-13) is used most often for electrical utilization equipment that requires a 20-amp branch circuit. This is so a 15A- or a 20A-cord can be plugged into it. Any receptacle devices that are 120 volt, 20A rated will have this feature.

Figure 6-13: Single and duplex receptacle

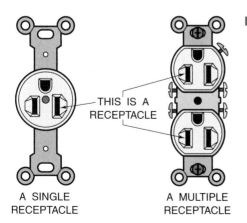

Figure 6-14: Ground fault circuit interrupter receptacles—GFCI

The *120-volt 15-amp duplex receptacle* (see Figure 6-13) is the most common receptacle in use today.

The *120-volt 15/20-amp duplex receptacle* is very similar to the previous receptacle except that it has the "T" slot to accommodate a 20-amp load.

See pages 95–96 for the procedure for installing duplex receptacles.

The *GFCI duplex receptacle* (see Figure 6-14) is designed to trip when there is a difference in the current going to the load and the current returning from the load. This device will trip if there is a difference of 4mA between the two. This device is easily recognizable because of the Trip and Reset buttons that are on the face of the receptacle. Once this receptacle is properly installed on a branch circuit, every device and fixture connected to the branch circuit past the GFCI receptacle is protected. This receptacle is to be used above countertops in bathrooms and kitchens, in wet or damp locations (such as basements), and outside. This receptacle is available in 15 and 20 amps as well.

See page 98 for the procedure for installing GFCI duplex receptacles.

Fixtures

There are many different types of lighting fixtures available today. Only the most common types will be discussed in this section.

Fixtures are described by the way they mount and by the type of bulb that is used within them. For example, a surface-mounted, incandescent, ceiling fixture is one that mounts against the ceiling surface and has an incandescent bulb. An incandescent bulb is the most commonly used light bulb. It is the standard frosted or clear light bulb that we use in our homes. It tends to give off a yellowish light when compared to a fluorescent light bulb. Fluorescent bulbs, most commonly referred to as "tubes," tend to give of a white light.

Here is a list of the most common light fixtures:

- Surface-mount incandescent fixture: mounts against ceiling or wall (see Figure 6-15 on page 99)
- Surface-mount fluorescent fixture: mounts against ceiling or wall (see Figure 6-16 on page 99)

Procedures

Installing Duplex Receptacles in a Nonmetallic Electrical Outlet Box

- Wear safety glasses and observe all applicable safety rules.

A Using a wire stripper, remove approximately 3/4 inch of insulation from the end of the insulated wires.

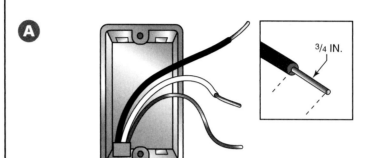

B Using long-nose pliers or wire strippers, make a loop at the end of each of the wires.

- Place the loop on the black wire around a brass terminal screw so that the loop is going in the clockwise direction. While pulling the loop snug around the screw terminal, tighten the screw to the proper amount with a screwdriver.

- Place the loop on the white wire around a silver terminal screw so that the loop is going in the clockwise direction. While pulling the loop snug around the screw terminal, tighten the screw to the proper amount with a screwdriver.

C Complete the installation by placing the loop on the bare grounding wire around the green terminal screw so that the loop is going in the clockwise direction. While pulling the loop snug around the screw terminal, tighten the screw to the proper amount with a screwdriver.

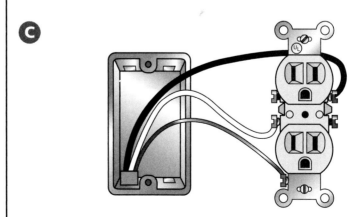

Procedures

Installing Duplex Receptacles in a Nonmetallic Electrical Outlet Box (Continued)

D Place the receptacle into the outlet box by carefully folding the conductors back into the device box.

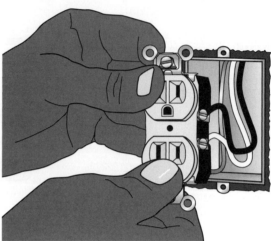

E Secure the receptacle to the device box using the 6-32 screws. Mount the receptacle so it is vertically aligned.

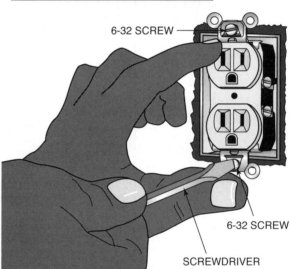

F Attach the receptacle cover plate to the receptacle. Be careful to not tighten the mounting screw(s) too much. Plastic faceplates tend to crack very easily.

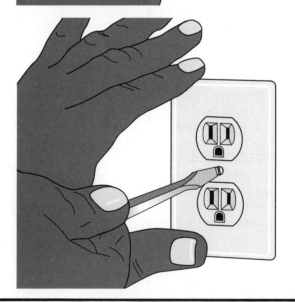

Procedures

Installing Duplex Receptacles in a Metal Electrical Outlet Box

- Wear safety glasses and observe all applicable safety rules.

 Attach a 6- to 8-inch-long grounding pigtail to the metal electrical outlet box with a 10-32 green grounding screw. The pigtail can be a bare or green insulated copper conductor.

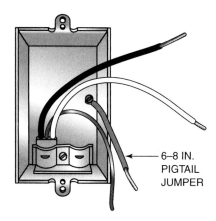

 Attach another 6- to 8-inch-long grounding pigtail to the green screw on the receptacle.

 Using a wirenut, connect the branch-circuit grounding conductor(s), the grounding pigtail attached to the box, and the grounding pigtail attached to the receptacle together.

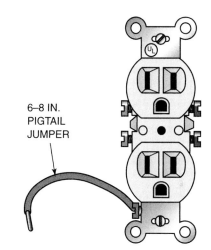

- Using a wire stripper, remove approximately 3/4 inch of insulation from the end of the insulated wires.

- Using long-nose pliers or wire strippers, make a loop at the end of each of the wires.

- Place the loop on the black wire around a brass terminal screw and the loop on the white wire around a silver terminal screw so that the loops are going in the clockwise direction. Tighten the screws to the proper amount with a screwdriver.

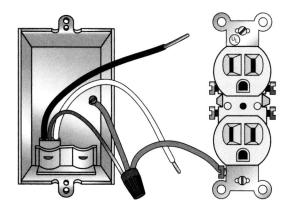

- Place the receptacle into the outlet box by carefully folding the conductors back into the device box.

- Secure the receptacle to the device box using the 6-32 screws. Mount the receptacle so it is vertically aligned.

- Attach the receptacle cover plate to the receptacle. Be careful to not tighten the mounting screw(s) too much. Plastic faceplates tend to crack very easily.

Procedures

Installing Feed-Through GFCI and AFCI Duplex Receptacles in Nonmetallic Electrical Outlet Boxes

- Wear safety glasses and observe all applicable safety rules.

- **A** At the electrical outlet box containing the GFCI or AFCI feed-through receptacle, use a wirenut to connect the branch-circuit grounding conductors and a grounding pigtail together. Connect the grounding pigtail to the receptacle's green grounding screw.

- At the electrical outlet box containing the GFCI or AFCI feed-through receptacle, identify the incoming power conductors and connect the white grounded wire to the line-side silver screw and the incoming black ungrounded wire to the line-side brass screw.

- At the electrical outlet box containing the GFCI or AFCI feed-through receptacle, identify the outgoing conductors and connect the white grounded wire to the load-side silver screw and the outgoing black ungrounded wire to the load-side brass screw.

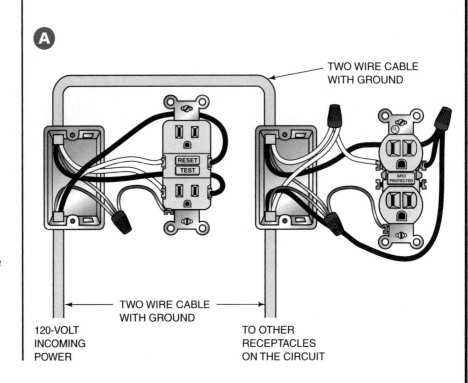

- Secure the GFCI or AFCI receptacle to the electrical box with the 6-32 screws provided by the manufacturer.

- A proper GFCI or AFCI cover is provided by the device manufacturer; attach it to the receptacle with the short 6-32 screws also provided.

- At the next "downstream" electrical outlet box containing a regular duplex receptacle, connect the white grounded wire(s) to the silver screw(s) and the black ungrounded wire(s) to the brass screw(s) in the usual way. Place a label on the receptacle that states "GFCI Protected." These labels are provided by the manufacturer.

- Continue to connect and label any other "downstream" duplex receptacles as outlined in the previous step.

- Recessed-can incandescent lighting fixture: mounts in the ceiling (see Figure 6-17)
- Pendant-type incandescent lighting fixture: hangs from the ceiling on a chain or cable (see Figure 6-18)
- Chandelier-type lighting fixture: multiple lamp fixtures hang from the ceiling on a chain or cable (see Figure 6-19)

Figure 6-15: Wall-mounted light fixture

Figure 6-16: Ceiling-mounted light fixture

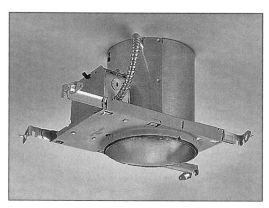

Figure 6-17: Recessed-can lighting fixture

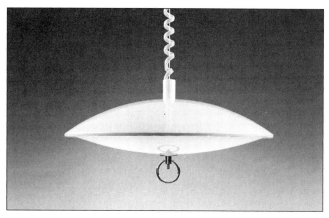

Figure 6-18: Pendant-type lighting fixtures

Figure 6-19: Chandelier-type lighting fixtures

Figure 6-19: (Continued)

Measurement Instruments

The most important thing to remember when working with electricity is to turn off the power. It is also important to learn how to make sure the power is off, and this can be done easily by using a variety of different testers.

Continuity Tester: used to indicate whether there is a continuous path for current flow in an electrical circuit or electrical device (see Figure 6-20).

Figure 6-20: Continuity tester

Note: Never attach a continuity tester to a circuit that is energized.

Voltage Testers and Voltmeters: used to verify that a voltage is present or measuring for a certain amount of voltage (see Figures 6-21, 6-22 and 6-23).

A noncontact voltage tester (see Figure 6-24) does not need to have contact with the circuit.

See pages 102–104 for the procedure for using a voltage tester.

Ammeter: used to measure the amount of current flowing in a circuit. They can also be used to balance the loads on multiwire circuits and to locate electrical component malfunctions.

The two types of ammeter (see Figure 6-25 on page 105):

- In-line: seldom used because they do not have high enough amperage ratings
- Clamp-on: most often used type of ammeter (see Figure 6-26 on page 105)

Note: Always read and follow the instructions that are supplied with the ammeter.

Ohmmeter and Megohmmeter: used to measure the resistance of a circuit or circuit component (see Figure 6-27)

See page 106 for the procedure for using a clamp-on ammeter.

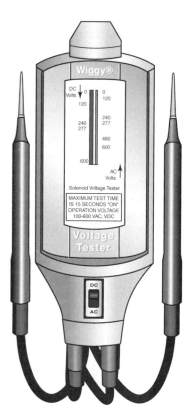

Figure 6-21: **Common style of voltage tester**

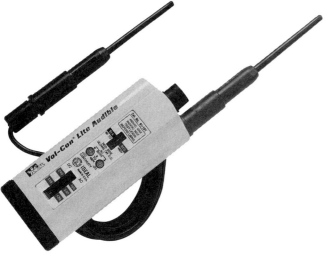

Figure 6-22: **Voltage tester and continuity tester combination**

Figure 6-23: **Digital voltage tester**

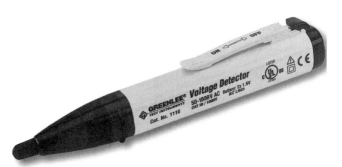

Figure 6-24: **Noncontact voltage tester**

Procedures

Using a Voltage Tester

- In this example, we will determine which conductor of a circuit is grounded using a voltage tester.

- Put on safety glasses and observe regular safety procedures.

A Connect the tester between one circuit conductor and a well-established ground.

- If the tester indicates a voltage, the conductor being tested is not grounded.

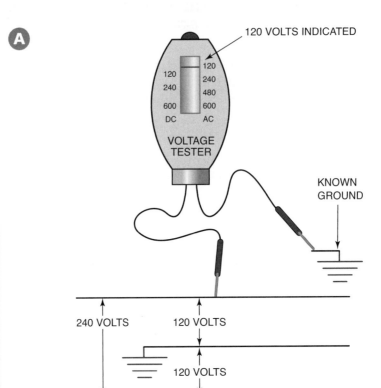

B Continue this procedure with each conductor until zero voltage is indicated between the tested conductor and the known ground. Zero voltage indicates that you have found a grounded circuit conductor.

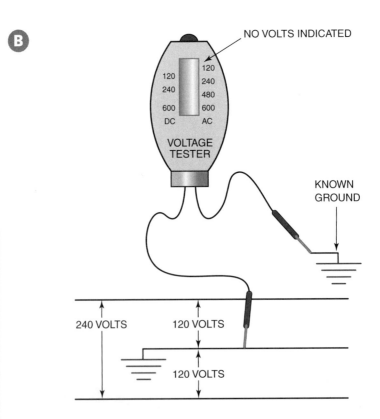

- In this example, we will determine the approximate voltage between two conductors using a voltage tester.

- Put on safety glasses and observe regular safety procedures.

- **C** Connect the tester between the two conductors.

- Read the indicated voltage value on the meter. Note: With a solenoid type tester you should also feel a vibration that is another indication of voltage being present.

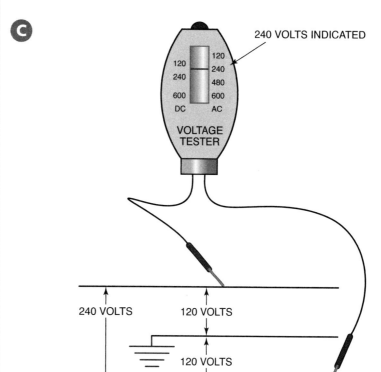

Procedures: Using a Noncontact Voltage Tester

- In this example, we will determine if an electrical conductor is energized using a non-contact voltage tester.

- Put on safety glasses and observe regular safety procedures.

- Identify the conductor to be tested.

- **(A)** Bring the noncontact voltage tester close to the conductor. Note that some noncontact voltage testers may have to be turned on before using.

- Listen for the audible alarm, observe a light coming on, or feel a vibration to indicate that the conductor is energized.

(A) THE TESTER WILL PROVIDE A VISUAL OR AUDIBLE INDICATION IF THERE IS A VOLTAGE PRESENT.

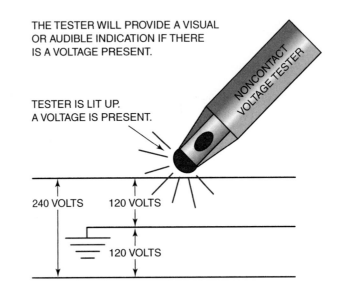

TESTER IS LIT UP. A VOLTAGE IS PRESENT.

240 VOLTS | 120 VOLTS | 120 VOLTS

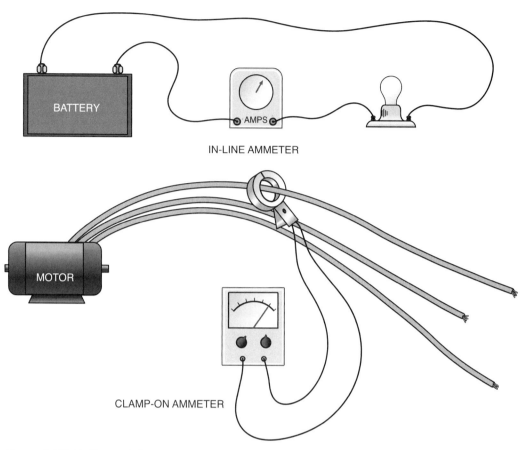

Figure 6-25: In-line and clamp-on ammeters

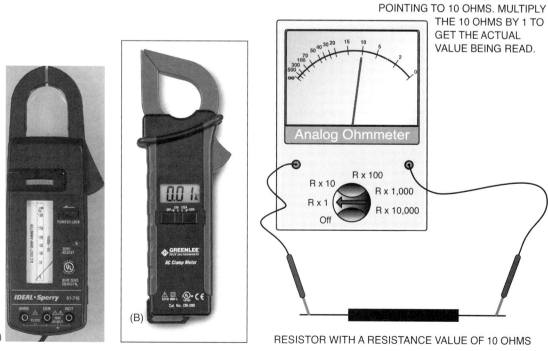

Figure 6-26: Analog and digital clamp-on ammeters

Figure 6-27: Analog ohmmeter

Procedures: Using a Clamp-On Ammeter

- In this example, we will be measuring current flow through a conductor with a clamp-on ammeter. Note that you can take a current reading with a clamp-on ammeter clamped around only one conductor. For example, a clamp-on meter will not give a reading when clamped around a two-wire Romex cable.

- Put on safety glasses and observe regular safety procedures.

- If the meter is analog and has a scale selector switch, set it to the highest scale. Skip this step if the meter is digital and has an auto-ranging feature.

 Open the clamping mechanism and clamp it around the conductor.

- Read the displayed value.

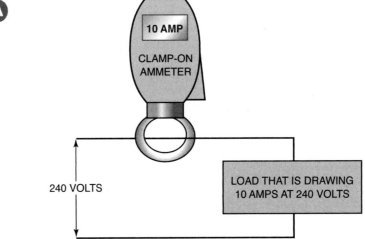

Electrical Maintenance Procedures

Troubleshooting

Troubleshooting is a process in which a person gathers information and forms a logical conclusion as to the problems that may be present within the system. All problems should be looked at logically. Take time and consider the most logical problems that would cause the symptoms that are present in the faulty circuit. If you do this, you have a higher chance of success in finding and fixing the problem. All problems that occur in a system will give tell-tale signs that will help you find the problem. Training is always beneficial. This gives you a knowledge base that will help you form the logical conclusions needed to solve problems with the system.

Troubleshooting occurs while you are gathering the information about the faulty system. Diagnosing begins while the data is being collected and is completed when a decision as to what the problem may be is formulated. Once you have diagnosed a problem, you must prove the diagnosis and repair the problem.

Below is a simplified, step-by-step guideline that could be used while troubleshooting. *Remember: Do not perform any of the following steps unless you are a qualified individual.*

1. You are notified of a problem.
2. Ask the person who has reported the problem as many questions as possible as to what was witnessed during the failure. This may include something that was seen, smelling a distinct or peculiar odor, or feeling heat in general vicinity of the problem.
3. Begin troubleshooting while making sure that all safety standards are adhered to. It is a good idea to start troubleshooting at the most logical area that would cause the described symptoms. You should start diagnosing the problem as soon as you receive the descriptions given to you and as you start receiving data from troubleshooting.
4. Safely remove any covers or panels that will give you access to the part of the electrical system that you are troubleshooting.
5. Visually inspect the equipment and devices for any signs of overheating or disintegration. If signs are visible, go to step 6a. If signs are not visible, go to step 6b below.
6a. *Be sure that you are qualified to work on the device and equipment before working on any part of an electrical system.* De-energize the circuit or system, attach your lock-out/tag-out device to the disconnecting means, and place the key in *your* pocket. Go to step 7.
6b. Take the appropriate step to safely acquire voltage and/or current readings at the suspected device or equipment. If your readings indicate that voltage is present and current is not following as it should be, the suspected device or equipment may be faulty and it may need to be replaced. If you decide to replace the faulty device or equipment, de-energize the circuit or system, attach your lock-out/tag-out device to the disconnecting means, and place the key in *your* pocket; then go to step 7.
7. Go back to the device or equipment and verify that the circuit is in fact deenergized.

8. Once it has been verified that the circuit is de-energized, begin working on your fault.

   **In the event that you may need to repair and/or replace common electrical devices such as receptacles, switches, interior and exterior lighting fixtures, bulbs, or ballasts, follow the simple procedures listed in the last section of this chapter.

9. Once the fault is repaired, and all covers are back in place, remove the lock-out/tag-out from the source of energy and re-energize the circuit.
10. Go back to the device or equipment that was replaced and verify that it is working properly. If it is, inform the person who called the job in. If it is not working, you may want to consider calling a qualified electrician to come troubleshoot and diagnose the problem.

Perform Tests

It is necessary to regularly perform tests on the following to ensure that they are operating properly before an emergency arises:

- smoke alarms
- fire alarms
- medical alert systems
- emergency exit lighting
- GFCI receptacles

Test Smoke Alarms and Fire Alarms

Individual smoke alarms and fire alarms typically have a test button that can be pushed. Be aware that pushing the test button on any one alarm may set off all alarms that are on the system as the fire alarm code requires all of them to be tied together.

Some fire alarm systems may require you to put the system in test mode before testing. If this is not done and the fire alarm is activated during the test, the sprinkler systems may activate. Placing the fire alarm system in test mode would allow the electrical portion of the system to be tested without the sprinklers activating during the test. Be sure to repair or replace any defective alarm or smoke detector according to manufacturer's specifications and report any malfunctions to your supervisor.

Many of the smoke detectors have a 9-volt battery in them so that the alarm can continue to work in the event that there is a loss of power. The batteries should be checked on a regular basis. If a detector is found to have a dead or weak battery, the battery should be changed immediately. If the smoke detector does not operate properly, it should be changed immediately (see Figure 6-28).

To test a smoke detector:

1. Press the battery-test button on the unit to make sure the battery is properly connected.
2. If the unit has a battery that's more than a year old, replace the battery.

Figure 6-28: **Smoke detector**

3. Light a candle and hold it approximately 6 inches below the detector so that heated air will rise into the unit.
4. If the alarm doesn't sound within 20 seconds, blow out the candle and let the smoke rise into the unit.
5. If the alarm still doesn't sound, open the unit up and make sure it is clean and that all electrical connections are solid.
6. If, again, the alarm doesn't sound, replace the smoke detector.

To replace a smoke detector battery:

1. Remove the smoke detector cover, typically by carefully pulling down on the case's perimeter or by twisting the case counterclockwise.
2. Locate and remove the battery.
3. Replace it with a new one.
4. Close the case and test the smoke detector.

Read the owner's manual for additional troubleshooting tips and possible adjustments.

Test GFCI Receptacle

1. Go to the receptacle and locate the "Test" button.
2. Press the test button and listen for a very light "pop" in the receptacle. You may also notice a small indicator that will light up after you have tripped the GFCI. If the GFCI tripped, then it is working correctly.
3. Press the "Reset" button and the GFCI should reset and the indicator (if preset) will turn off.

If it did not trip:

4. Press the test button again.

If the GFCI still does not trip, then it is faulty and needs to be replaced, or it could be wired incorrectly and therefore would need to be rewired correctly. For the correct installation procedures of a GFCI receptacle, see page 114.

Test Medical Alert Systems

Assisted living facilities may have and hospitals will have medical alert systems. If your facility has a medical alert system, tests that are conducted on these systems should be performed only by qualified personnel who have been properly trained on the proper test procedures for that system. All tests that are performed on a medical alert system must be performed according to the manufacturer's specifications. If any defects are found in the system, they should be reported to your supervisory immediately.

Replacing Detectors, Devices, Fixtures, and Bulbs

Be sure that as you replace or repair any defective equipment or devices, that you do so according to manufacturer's specifications. Also, don't forget to report any malfunctions to your supervisor.

Replacing a Smoke Detector

1. Acquire a fiberglass or non-conductive ladder that will be tall enough for you to reach the detector once you are on it. DO NOT USE AN ALUMINUM LADDER. Set the ladder up under the faulty detector, making sure that all four legs of the ladder are solidly in place.
2. Go to the panel and de-energize the branch circuit that supplies the smoke detectors.
3. Lock-out/tag-out the breaker (see Figure 6-29).
4. Go back to the faulty detector and remove it from its mounting base by twisting the detector body in a counterclockwise direction. The detector should release from the base (see Figure 6-30).
5. Make a note of how the smoke detector is wired. If necessary, make a sketch on a piece of paper to follow when you are re-connecting the power leads.

You may find that the smoke detector you are replacing has a quick connector plug on the back of the smoke detector. If you are replacing the faulty detector with a new one that is the same model, simply unplug the quick connect from the smoke detector. If a quick connect is not present, you may have to disconnect the power leads from the smoke detector by removing the wirenuts. There should be three leads on a newer type of smoke detector (see Figure 6-31). One will be black, one will be white, and one will be orange. The black wire is the hot wire, the white wire is the neutral, and the orange wire is for the repeating circuit. (This is the wire that triggers all of the other smoke alarms.)

Figure 6-29: **Examples of lock-out/tag-out devices**

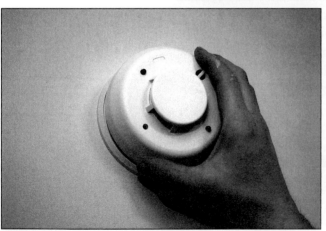

Figure 6-30: **Removing faulty detector**

![Figure 6-31 diagram of interconnected smoke detectors with labels: 120-VOLT CIRCUIT, OUTLET BOXES IN CEILING, CEILING, PLUG-IN POWER BLOCK ON REAR OF ALARM, SMOKE ALARMS, BATTERY COMPARTMENT]

Figure 6-31: **Interconnected smoke detectors**

6. If you are replacing the faulty detector with a new one that is the same model, you may be able to leave the mounting base mounted to the ceiling. If you are replacing the faulty smoke detector with a different brand or model, you may find that the base of the old one will not work on the new one, and that the old mounting base must therefore be removed. This can be accomplished by loosening the two screws that are holding the base to the box. Once the base is loose, simply twist it in a counterclockwise direction as you did with the detector. The mounting base should release from the box. Now replace the old base with the new base and tighten the screws (see Figure 6-32).

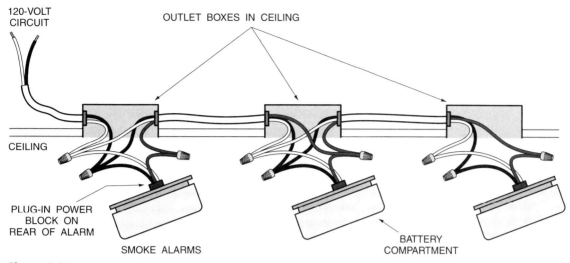

Figure 6-32: **Mounting base still attached to box**

7. Reconnect the power leads. If you are replacing an older two-wire model with a newer three-wire model, you will not need to use the orange lead. Simply cap the orange lead and stuff it into the box.
8. Set the new detector up against the mounting base and twist it in a clockwise direction until it is securely mounted.
9. Go to the panel, remove your lock-out from the breaker, and turn on the breaker. If the breaker does not trip, go to the next step. If the breaker trips, set the breaker to the off position and lock it out again. Go back to the detector, remove it from the base, and visually inspect the connections. Be sure that there is no exposed copper wire on the black lead that could be touching the bare copper ground wire. Once you think everything is correct, put the detector back on its base, go back to the panel, remove your lock-out from the breaker, and turn the breaker back on. If the breaker trips again, call a qualified electrician to come and troubleshoot the problem. If the breaker does not trip, go to the next step.
10. Go back to the detector and verify that it is working correctly.
11. Gather all tools and ladders, and clean up the area that you were working in.

Replacing a Switch

Remove the old switch:

1. Shut off the power at the circuit breaker box.
2. Remove the cover plate (see Figures 6-33 and 6-34) and test the terminals with the circuit tester. If the tester does not light up, there is no power going to the switch.
3. Remove the screws that hold the switch in place and pull the switch from the wall.
4. Loosen the screws holding the wires to the switch, remove the wires, and remove the switch. In some newer switches, the wires may go directly into the switch, where they are held in place by clamps inside the switch. These switches usually have a slot into which you can insert a small screwdriver to loosen the clamps.

Figure 6-33: Cover plate (©2006 JupiterImages Corporation)

Figure 6-34: Cover plate removed (©2006 JupiterImages Corporation)

Figure 6-35: Replacement switch (©2006 JupiterImages Corporation)

Figure 6-36: New switch (©2006 JupiterImages Corporation)

Install the new switch:

1. Begin by bending the end of the ground wire into a small hook, placing the hook over the ground terminal of the switch, and tightening it into place (see Figure 6-35).
2. Attach the remaining wires to the terminals and tighten them in the same way. Your new switch may have holes that allow you to insert the wires without using the terminal screws. If so, straighten the wires, press them into the holes as far as they will go, and then tug on them to make sure they are held securely in place.
3. Gently press the switch back into position and secure it in place with screws. Then replace the switch cover (see Figure 6-36) and turn on the power at the circuit breaker box.

Replacing a Receptacle

The following steps should be used to replace an existing receptacle.

1. Turn the power off to the receptacle.
2. Test the receptacle to ensure that the power is off.
3. Remove the cover plate and receptacle from the box.
4. Mark the common wire with a piece of tape. The common wire terminal of the receptacle will be a different color that the other terminals.

5. Detach the common wire from the old receptacle, followed by the neutral and ground wires.
6. Locate the common terminal on the new receptacle and attach the wire. Remember that the common terminal will be a different color.
7. Attach the neutral and ground wire to the new receptacle.
8. Push the receptacle back into the box.
9. Install the receptacle cover onto the new receptacle.
10. Turn the power back on.

Replacing a GFCI Receptacle

A GFCI receptacle has a line side and a load side (see Figure 6-37).

If the GFCI receptacle has only one black and one white wire terminated to it (not counting the ground wire), then both of the wires are to be terminated to the line side of the GFCI receptacle while making sure to connect the black wire to the brass colored screw, the white wire to the silver screw, and the bare wire (if present) to the green colored screw.

If other receptacles on the branch circuit are fed through the GFCI receptacle, there will be five wires on the receptacle. Two black wires, two white wires, and one ground wire. If this is the case, it is important to terminate the wires that are feeding the line voltage into the GFCI receptacle to the line side of the GFCI receptacle. The wires that are feeding the remaining receptacles on the branch circuit should be connected to the load side of the GFCI receptacle (see the procedure on page 114).

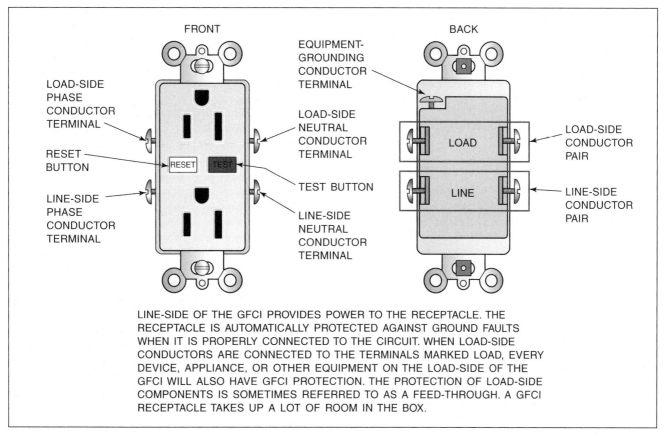

Figure 6-37: **Ground Fault Circuit Interrupter (GFCI)**

Procedures

Installing Feed-Through GFCI and AFCI Duplex Receptacles in Nonmetallic Electrical Outlet Boxes

- Wear safety glasses and observe all applicable safety rules.

- **A** At the electrical outlet box containing the GFCI or AFCI feed-through receptacle, use a wirenut to connect the branch-circuit grounding conductors and a grounding pigtail together. Connect the grounding pigtail to the receptacle's green grounding screw.

- At the electrical outlet box containing the GFCI or AFCI feed-through receptacle, identify the incoming power conductors and connect the white grounded wire to the line-side silver screw and the incoming black ungrounded wire to the line-side brass screw.

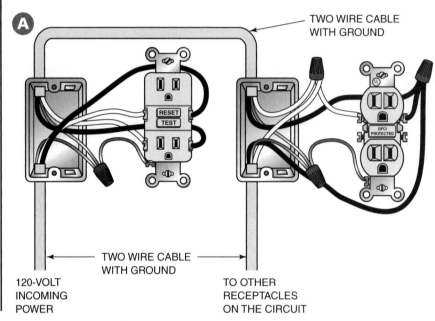

- At the electrical outlet box containing the GFCI or AFCI feed-through receptacle, identify the outgoing conductors and connect the white grounded wire to the load-side silver screw and the outgoing black ungrounded wire to the load-side brass screw.

- Secure the GFCI or AFCI receptacle to the electrical box with the 6-32 screws provided by the manufacturer.

- A proper GFCI or AFCI cover is provided by the device manufacturer; attach it to the receptacle with the short 6-32 screws also provided.

- At the next "downstream" electrical outlet box containing a regular duplex receptacle, connect the white grounded wire(s) to the silver screw(s) and the black ungrounded wire(s) to the brass screw(s) in the usual way. Place a label on the receptacle that states "GFCI Protected." These labels are provided by the manufacturer.

- Continue to connect and label any other "downstream" duplex receptacles as outlined in the previous step.

Procedures: Installing a Light Fixture Directly to an Outlet Box

- Put on safety glasses and observe all applicable safety rules.

- Using a voltage tester, verify that there is no electrical power at the lighting outlet where the fixture will be installed. If electrical power is present, turn off the power and lock out the circuit.

- Locate and identify the ungrounded, grounded, and grounding conductors in the lighting outlet box.

A The grounding conductor will not be connected to this fixture. If there is a grounding conductor in a nonmetallic box, simply coil it up and push it to the back or bottom of the electrical box. Do not cut it off, as it may be needed if another type of light fixture is installed at that location. If there are two or more grounding conductors in a nonmetallic box, connect them to-

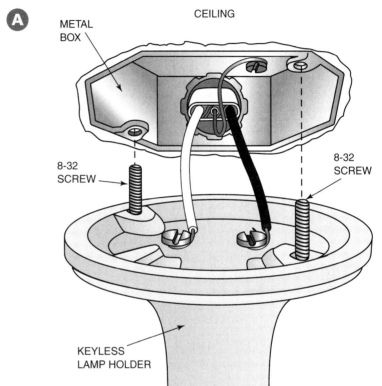

gether with a wirenut and push them to the back or bottom of the lighting outlet box. If there is a grounding conductor in a metal outlet box, it must be connected to the outlet box by means of a listed grounding screw or clip. If there are two or more grounding conductors in a metal box, use a wirenut to connect them together along with a grounding pigtail. Attach the grounding pigtail to the metal box with a listed grounding screw.

- Connect the white grounded conductor(s) to the silver screw or white fixture pigtail. If there is one grounded conductor in the box, strip approximately 3/4 inch (19 mm) of insulation from the end of the conductor and form a loop at the end of the conductor using an approved tool, such as a T-stripper. Once the loop is made, slide it around the silver terminal screw on the fixture so the end is pointing in a clockwise direction. Hold the conductor in place and tighten the screw. If there are two or more grounded conductors in the box, strip the ends as described previously and use a wirenut to connect them and a white pigtail together. Attach the pigtail to the silver grounded screw as described previously.

- The black ungrounded conductor(s) is connected to the brass-colored terminal screw on the fixture. The connection procedure is the same as for the grounded conductor(s).

- Now the fixture is ready to be attached to the lighting outlet box. Make sure the grounding conductors are positioned so they will not come in contact with the grounded or ungrounded screw terminals. Align the mounting holes on the fixture with the mounting holes on the lighting outlet box. Insert the 8-32 screws that are usually equipped with the fixture through the fixture holes. Then thread them into the mounting holes in the outlet box.

- Tighten the screws until the fixture makes contact with the ceiling or wall. Be careful not to overtighten the screws, as you may damage the fixture.

- Install the proper lamp, remembering not to exceed the recommended wattage.

- Turn on the power and test the light fixture.

Procedures: Installing a Cable-Connected Fluorescent Lighting Fixture Directly to the Ceiling

- Put on safety glasses and observe all applicable safety rules.

- Using a voltage tester, verify that there is no electrical power at the lighting outlet where the fixture will be installed. If electrical power is present, turn off the power and lock out the circuit.

- Place the fixture on the ceiling in the correct position, making sure it is aligned and the electrical conductors have a clear path into the fixture.

- Mark on the ceiling the location of the mounting holes.

- Use a stud finder to determine if the mounting holes line up with the ceiling trusses. If they do, screws will be used to mount the fixture. If they do not, toggle bolts will be necessary. For some installations, a combination of screws and toggle bolts will be required.

- If screws are used, drill holes into the ceiling using a drill bit that has a smaller diameter than the screws to be used. This will make installing the screws easier. If toggle bolts are to be used, use a flat-bladed screwdriver to punch a hole in the sheetrock only large enough for the toggle to fit through.

- Remove a knockout from the fixture where you wish the conductors to come through. Install a cable connector in the knockout hole.

- Place the fixture in its correct position and pull the cable through the connector and into the fixture. Tighten the cable connector to secure the cable to the fixture. This part of the process may require the assistance of a coworker. If using toggle bolts, put the bolt through the mounting hole and start the toggle on the end of the bolt.

- With a coworker holding the fixture, install the mounting screws or push the toggle through the hole until the wings spring open. This will hold the fixture in place until the fixture is secured to the ceiling.

- Make the necessary electrical connections. The grounding conductor should be properly wrapped around the fixture grounding screw and the screw tightened. The white grounded conductor is connected to the white conductor lead, then the black ungrounded conductor is connected to the black fixture conductor.

- Install the wiring cover by placing one side in the mounting clips, squeezing it, and then snapping the other side into its mounting clips.

- Install the recommended lamps. Usually, they have two contact pins on each end of the lamp. Align the pins vertically, slide them up into the lamp holders at each end of the fixture, and rotate the lamp until it snaps into place.

- Test the fixture and lamps for proper operation.

- Install the fixture lens cover.

Procedures

Installing a Strap on a Lighting Outlet Box Lighting Fixture

- Put on safety glasses and observe all applicable safety rules.

- Using a voltage tester, verify that there is no electrical power at the lighting outlet where the fixture will be installed. If electrical power is present, turn off the power and lock out the circuit.

- Before starting the installation process, read and understand the manufacturer's instructions.

- **A** Mount the strap to the outlet box using the slots in the strap. With metal boxes, the screws are provided with the box. With non-metallic boxes, you must provide your own 8-32 mounting screws. Put the 8-32 screws through the slot and thread them into the mounting holes on the outlet box. Tighten the screws to secure the strap to the box.

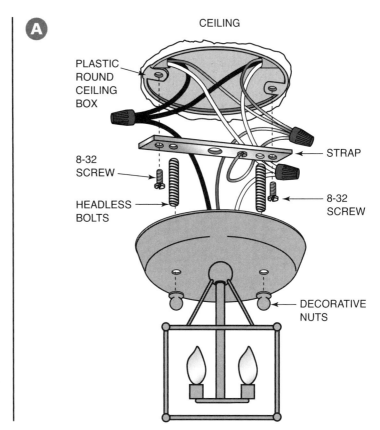

- Identify the proper threaded holes on the strap and install the fixture-mounting headless bolts in the holes so the end of the screw will point down.

- Make the necessary electrical connections. Make sure that *all* metal parts (including the outlet box), the strap, and the fixture are properly connected to the grounding conductor in the power feed cable.

- Neatly fold the conductors into the outlet box. Align the headless bolts with the mounting holes on the fixture. Slide the fixture over the headless bolts until the screws stick out through the holes. Do not be alarmed if the mounting screws seem to be too long. Thread the provided decorative nuts onto the headless bolts. Keep turning the nuts until the fixture is secure to the ceiling or wall.

- Install the recommended lamp and test the fixture operation.

- Install any provided lens or globe. They are usually held in place by three screws that thread into the fixture. Start the screws into the threaded holes, position the lens or globe so it touches the fixture, and tighten the screws until the globe or lens is snug. Do *not* over tighten the screws. You may return the next day and find the globe or lens cracked or broken.

Procedures

Installing a Chandelier-Type Light Fixture Using the Stud and Strap Connection to a Lighting Outlet Box

- Put on safety glasses and observe all applicable safety rules.

- Using a voltage tester, verify that there is no electrical power at the lighting outlet where the fixture will be installed. If electrical power is present, turn off the power and lock out the circuit.

- Before starting the installation process, read and understand the manufacturer's instructions.

A Install the mounting strap to the outlet box using 8-32 screws.

- Thread the stud into the threaded hole in the center of the mounting strap. Make sure that enough of the stud is screwed into the strap to make a good secure connection.

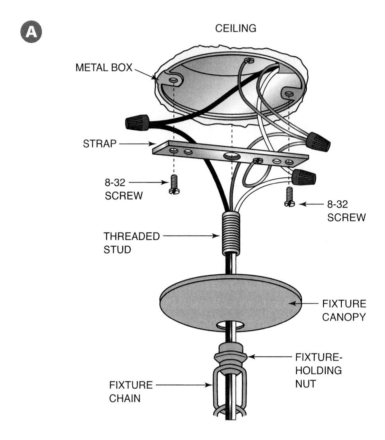

- Measure the chandelier chain for the proper length, remove any unneeded links and install one end to the light fixture.

- Thread the light fixture's chain-mounting bracket on to the stud. Remove the holding nut and slide it over the chain.

- Slide the canopy over the chain.

- Attach the free end of the chain to the chain-mounting bracket.

- Weave the fixture wires and the grounding conductor up through the chain links, being careful to keep the chain links straight. Section 410.28(F) of the *NEC®* states that the conductors must not bear the weight of the fixture. As long as the chain is straight and the conductors make all the bends, the chain will support the fixture properly.

- Now run the fixture wires up through the fixture stud and into the lighting outlet box.

- Make all necessary electrical connections.

- Slide the canopy up the chain until it is in the proper position. Slide the nut up the chain and thread it on to the chain-mounting bracket until the canopy is secure.

- Install the recommended lamp and test the fixture for proper operation.

Procedures

Installing a Fluorescent Fixture (Troffer) in a Dropped Ceiling

- Put on safety glasses and observe all applicable safety rules.

- Before starting the installation process, read and understand the manufacturer's instructions.

- During the rough-in stage, mark the location of the fixtures on the ceiling.

- Using standard wiring methods, place lighting outlet boxes on the ceiling near the marked fixture locations and connect them to the lighting branch circuit.

- Once the dropped ceiling grid has been installed by the ceiling contractor, install the fluorescent light fixtures in the ceiling grid at the proper locations. Some electricians refer to this action as "laying in" the fixture. Once the fixture is installed, some electricians refer to the fixtures as being "laid in."

 Support the fixture according to *NEC®* requirements. Section 410.16(C) requires that all framing members used to support the ceiling grid be securely fastened to each other and to the building itself. The fixtures themselves must be securely fastened to the grid by an approved means, such as bolts, screws, rivets, or clips. This is to prevent the fixture from falling and injuring someone.

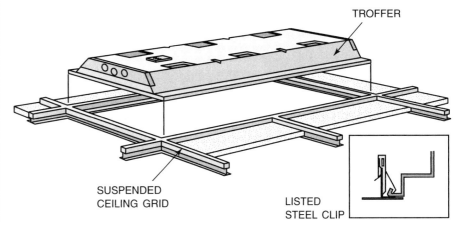

IMPORTANT: TO PREVENT THE LUMINAIRE (FIXTURE) FROM INADVERTENTLY FALLING, *410.16(C)* OF THE CODE REQUIRES THAT (1) SUSPENDED CEILING FRAMING MEMBERS THAT SUPPORT RECESSED LUMINAIRES (FIXTURES) MUST BE SECURELY FASTENED TO EACH OTHER, AND MUST BE SECURELY ATTACHED TO THE BUILDING STRUCTURE AT APPPROPRIATE INTERVALS, AND (2) RECESSED LUMINAIRES (FIXTURES) MUST BE SECURELY FASTENED TO THE SUSPENDED CEILING FRAMING MEMBERS BY BOLTS, SCREWS, RIVETS, OR SPECIAL LISTED CLIPS PROVIDED BY THE MANUFACTURER OF THE LUMINAIRE (FIXTURE) FOR THE PURPOSE OF ATTACHING THE LUMINAIRE (FIXTURE) TO THE FRAMING MEMBER.

Procedures

Installing a Fluorescent Fixture (Troffer) in a Dropped Ceiling (Continued)

- Using a voltage tester, verify that there is no electrical power at the lighting outlet where the fixture will be installed. If electrical power is present, turn off the power and lock out the circuit.

- **B** Connect the fixture to the electrical system. This is done by means of a "fixture whip." A fixture whip is often a length of Type NM, Type AC, or Type MC cable. It can also be a raceway with approved conductors, such as flexible metal conduit or electrical nonmetallic tubing. The fixture whip must be at least 18 inches (450 mm) long and no longer than 6 feet (1.8 m).

- Make all necessary electrical connections. The fixture whip should already be connected to the outlet box mounted in the ceiling. Using an approved connector, connect the cable or raceway to the fixture outlet box and run the conductors into the outlet box. Make sure that all metal parts are properly connected to the grounding system. Connect the white grounded conductors together and then the black ungrounded conductors together. Close the connection box.

- Install the recommended lamps and test the fixture for proper operation.

- Install the lens on the fixture.

B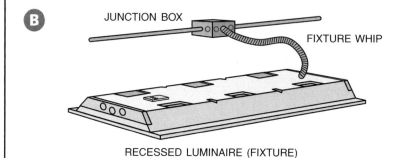

JUNCTION BOX

FIXTURE WHIP

RECESSED LUMINAIRE (FIXTURE)

Replacing a Light Fixture

Repair or Replace a 120-Volt Outlet

1. Turn off both the branch circuit and the main power at the service panel. Work in the daytime so that you can use natural light to light the work area. Also use a flashlight if needed.
2. Remove the plate and the outlet mounting screws (see Figure 6-38).
3. Pull outlet with wires still attached about 4 to 6 inches out of the junction box.
4. Note the color of the wires and identify the hot, neutral ground, and device ground.
5. Unscrew the terminal screws that attach the wire to the outlet and remove the wire. Start with the hot, then the neutral, and finally the ground.
6. Examine the new outlet. Identify which wire connects to which terminal. It does not matter which set of vertical screws you attach the wire to. If the outlet does not have markings indicating the polarity, then remember that the bright brass screw connects to your hot wire.
7. Using needle-nose pliers, connect the ground to the green terminal on the bottom of the outlet. Then connect the white wire to the neutral or silver terminal, and finally the hot wire to the hot or brass terminal. The wire should be wrapped completely around the terminal screws.
8. Finally, tuck the wires back in the junction box and mount the outlet and the outlet plate.

Figure 6-38: **Face plate**

Note: Make sure the outlet is rated for the amperage of the circuit. Do not use an outlet rated at 15 amps on a 20-amp circuit.

Replace Light Bulb

1. Turn off the lamp or light fixture.
2. Allow a hot bulb to cool before touching it.
3. Grasp the bulb lightly but firmly and turn counterclockwise until it is released from the socket (see Figure 6-39).
4. Insert a replacement bulb lightly but firmly into the socket and turn it clockwise until it is snug.
5. Turn the lamp or fixture back on.
6. Dispose of the used bulb.

Figure 6-39: **Remove light bulb**

Replace Fluorescent Bulb

1. Make sure the light switch is turned off.
2. Remove the lens or diffuser to access the bulb (see Figure 6-40). On most fluorescent lights, the lens or diffuser is a plastic panel below the bulb (see Figure 6-41). Push the panel up and tilt to remove.

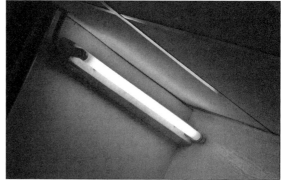
Figure 6-40: **Fluorescent light bulb** (©2006 Jupiter-Images Corporation)

Figure 6.41: Lens/diffuser on a fluorescent fixture

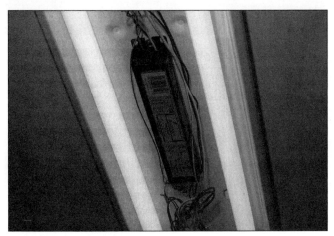

Figure 6-42: Ballast on a fluorescent fixture

3. Check to make sure that the problem is not something as simple as a poor contact. This can usually be corrected by giving the bulb a gentle turn a few degrees and then back to the lock position.
4. Hold the old bulb firmly at one end and rotate it one quarter-turn clockwise. This should put the end prongs in line with the loading slot.
5. Slide the bulb free.
6. Lower the end of the bulb carefully out of the socket. When one end is free, pull slightly and the other end should come out also.
7. Set the old bulb aside and lift a new bulb into the fixture.
8. Hold the bulb horizontally and rotate the new bulb until the prongs on each end are lined up with the grooves in the socket.
9. Insert the prongs in the socket and rotate the bulb a quarter-turn in a counterclockwise direction. The bulb should click into place on each end.
10. Test the light at the switch. If the light still doesn't come on, you may need to replace the ballast.

Replace Ballast

1. Find the ballast, which is usually seated near one end of the bulb. It is a silver, cylinder-shaped item with a diameter similar to that of a quarter. The ballast provides the starting voltage and then stabilizes the current for the fluorescent bulb(see Figure 6-42).
2. Loosen the ballast by turning it $\frac{1}{4}$- to $\frac{1}{2}$-turn counterclockwise.
3. Pull the ballast out of the fixture. Take it to a hardware store to match it with a new one.
4. Insert a new ballast into the light fixture; twist clockwise to lock into position.
5. Replace the lens cover.

Test and Replace Fuses

Identify and Replace Blown Fuses

1. Open the door to your service panel and examine it to locate the blown fuse. For light and receptacle circuits, look for a break or blackened area visible through the glass of a screw-in plug fuse (see Figure 6-43). If all the fuses look good, identify the fuse according to the circuit map printed on the door of the service panel or next to each fuse.
2. If the circuits are not mapped, locate the fuse by trial and error. Remove the fuses one at a time and insert a new fuse to test the circuit. You can also test the fuse with a continuity tester. Touch the pointed probe of a continuity tester to the fuse's tip and the clip to its threaded shaft. If the tester does not glow, the fuse is bad.

Figure 6-43: Screw in fuse (©2006 JupiterImages Corporation)

Figure 6-44: Fuse block

3. For fuse blocks (see Figure 6-44), which protect an electric stove and the main circuit, pull straight out on the handle and remove the individual cartridge fuses from the block by using a cartridge-fuse puller. Test the fuses with a continuity tester by probing the two ends.

Install a New Fuse

1. Screw in a new plug fuse, or install a new cartridge fuse in the fuse block and press the block back in. The replacement should always have the same rating as the original.
2. If all of the circuits have stopped working, remove and test the cartridge fuses in the main fuse block, usually located at the top left. (Occasionally it is reversed with the stove circuit on the top right.) Replace any faulty ones.

Reset Circuit Breaker

1. Push the breaker switch to the Off position and then back fully to the On position.
2. There will be a click as it snaps into the On position.

If the breaker trips again, you need to determine the reason for the overcurrent condition and correct the root cause of the problem. The breaker may be tripping due to excessive amperage in the circuit or may be shorting out.

Review Questions

1. What is an emergency backup system used for?

2. What is NM cable?

3. What is a single-pole switch?

4. What is a double-pole switch?

5. How do three-way switches differ from a single-pole switch?

6. What is a GFCI?

7. What is a continuity tester?

8. How is a smoke detector tested?

9. How are GFCI tested?

10. List the steps for replacing a smoke detector.

Name: _____

Date: _____

Electrical Facilities Maintenance Job Sheet 1

Electrical Facilities Maintenance

Electrical Troubleshooting and Maintenance

Upon completion of this job sheet, you should be able to demonstrate your ability to perform basic maintenance and electrical troubleshooting.

1. Choose an area in your facility and take an inventory of the switches being used in the area. What types are being used and what are they made of?

   _____
   _____

2. List five types of receptacles and what they are used for.

   _____
   _____

3. What is the most important thing to remember when working with electricity?

   _____
   _____

4. Define troubleshooting.

   _____
   _____

5. What is a GCFI receptacle and what is its purpose?

   _____
   _____

6. What is the purpose of a lock-out/tag-out device?

   _____
   _____

Instructor's Response:

_____
_____
_____

Chapter 7 Carpentry

OBJECTIVES

By the end of this chapter, you will be able to:

Knowledge-Based

- Describe the general properties of hardwood and softwood commonly used by facilities maintenance technicians.
- Describe the effects of moisture content on different wood products.
- Perform estimating and take-off quantities for simple one-step carpentry projects.
- Correctly identify and select engineered products, panels, and sheet goods.
- Correctly identify framing components.

Skill-Based

- Perform interior carpentry maintenance.
- Perform exterior carpentry maintenance.

Introduction

General Properties of Hardwood and Softwood

The carpenter works with wood more than any other material and must understand its characteristics in order to use it intelligently. Wood is a remarkable substance and is classified as either hardwood or softwood. There are different methods of classifying these woods. The most common method of classifying wood is by its source (see Figure 7-1). **Hardwood** comes from deciduous trees that shed their leaves each year. **Softwood** comes from coniferous, or cone-bearing, trees commonly know as evergreens.

Common hardwoods include:

- ash
- birch
- cherry
- hickory
- maple
- mahogany
- oak
- walnut

Figure 7-1: **Softwood and hardwood**

Common softwoods include:

- pine
- fir
- hemlock
- spruce
- cedar
- cypress
- redwood

The best way to learn the different kinds of woods is to work with them and examine them.

- Look at the color and the grain.
- Feel if it is heavy or light.
- Feel if it is hard or soft.
- Smell it for a characteristic odor.

Effects of Moisture Content

When a tree is first cut down, it contains a great amount of water. Lumber, when first cut from the log, is called **green lumber** and is very heavy because most of its weight is water (see Figure 7-2). A piece 2 inches thick, 6 inches wide, and 10 feet long may contain as much as 4¼ gallons of water, weighing about 35 pounds.

Green lumber should not be used in construction. As green lumber dries, it shrinks considerably unequally as the large amount of water leaves it. When it shrinks, it usually warps, depending on the way it was cut from the log (see Figure 7-3).

Realizing that lumber undergoes certain changes when moisture is absorbed or lost, the experienced carpenter uses techniques to deal with this characteristic of wood (see Figure 7-4).

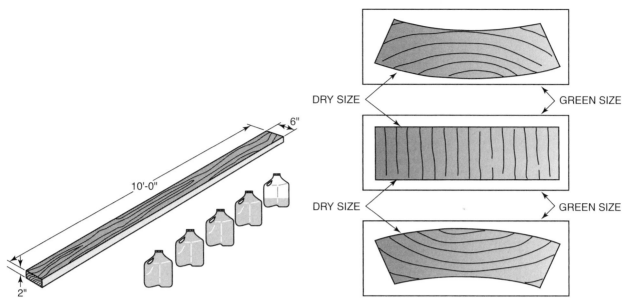

Figure 7-2: Green lumber contains a large amount of water

Figure 7-3: Lumber shrinks in the direction of the annular rings

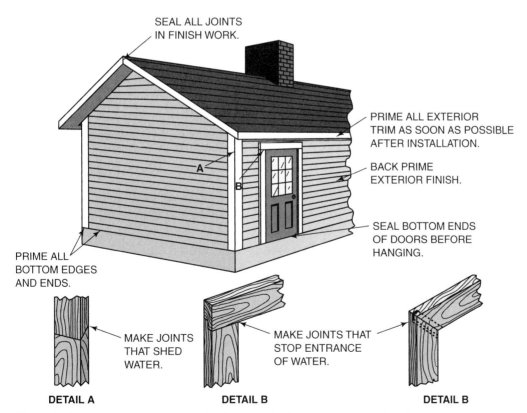

Figure 7-4: Techniques to prevent water from getting in behind the wood surface

Figure 7-5: APA performance-rated panels

Figure 7-6: Oriented strand board—OSB

Correctly Identify and Select Engineered Products, Panels, and Sheet Goods

Engineered Panels

The term **engineered panels** refers to man-made products in the form of large reconstituted wood sheets, sometimes called panels or boards.

- Plywood—one of the most extensively used engineered panels (Figure 7-5). Plywood is a sandwich of wood. Most plywood panels are made up of sheets of veneer called **plies**.
- Oriented strand board (OSB)—a non-veneered, performance-rated structural panel composed of small oriented (lined up) strand-like wood pieces arranged in three to five layers with each layer at right angles to the other (Figure 7-6).

Nonstructural Panels

Figure 7-7: Particleboard is made from wood flakes, shaving, resins, and waxes

- Hardwood plywood—available with hardwood face veneers, of which the most popular are birch, oak, and lauan.
- Particleboard—reconstituted wood panels made of wood flakes, chips, sawdust, and planer shavings (Figure 7-7). These wood particles are mixed with an adhesive, formed into a mat, and pressed into sheet form.
- Fiberboards—manufactured as high-density, medium-density, and low-density boards. Medium-density fiberboard (MDF) is manufactured in a manner similar to that used to make hardboard except that the fibers are not pressed as tightly. Low-density fiberboard is called **softboard**. Softboard is light and contains many tiny air spaces because the particles are not compressed tightly.

- Hardboards are high-density fiberboards. They are sometimes known by the trademark Masonite.

Other

- Gypsum board—used extensively in construction. Sometimes called wallboard, plasterboard, drywall, or the brand name Sheetrock (Figure 7-8). Gypsum board is readily available; easy to apply, decorate, or repair; and relatively inexpensive.
- Plastic laminates—used for surfacing kitchen cabinets and countertops. They are also used to cover walls or parts of walls in kitchens, bathrooms, and similar areas where a durable, easy-to-clean surface is desired.

Estimating

To estimate the amount of drywall material needed:

1. Determine the area of the walls and ceiling to be covered. Ceiling—multiply the length of the room by its width. Wall—multiply the perimeter of the room by the height.
2. Subtract only the large wall openings, such as double doors.
3. Combine all areas to find the total number of square feet of drywall.
4. Add about 5 percent of the total for waste.
5. Divide the total area to be covered by the area of one panel to get the number of panels.

Figure 7-8: **Drywall**

Framing Components

Wall Framing Components

The wall frame consists of a number of different parts. An exterior wall frame consists of the following components:

- Plates—top and bottom horizontal members of a wall frame.
- Studs—vertical members of the wall frame.
- Headers—run at right angles to studs (Figure 7-9).

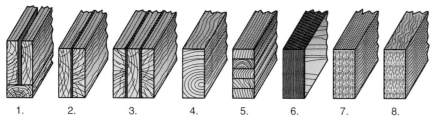

Figure 7-9: **Types of solid and built-up headers**

1. A BUILT-UP HEADER WITH A 2 X 4 OR 2 X 6 LAID FLAT ON THE BOTTOM.
2. A BUILT-UP HEADER WITH A $1/2$" SPACER SANDWICHED IN BETWEEN.
3. A BUILT-UP HEADER FOR A 6" WALL.
4. A HEADER OF SOLID SAWN LUMBER.
5. GLULAM BEAMS ARE OFTEN USED FOR HEADERS.
6. A BUILT-UP HEADER OF LAMINATED VENEER LUMBER.
7. PARALLEL STRAND LUMBER MAKES EXCELLENT HEADERS.
8. LAMINATED STRAND LUMBER IS USED FOR LIGHT DUTY HEADERS.

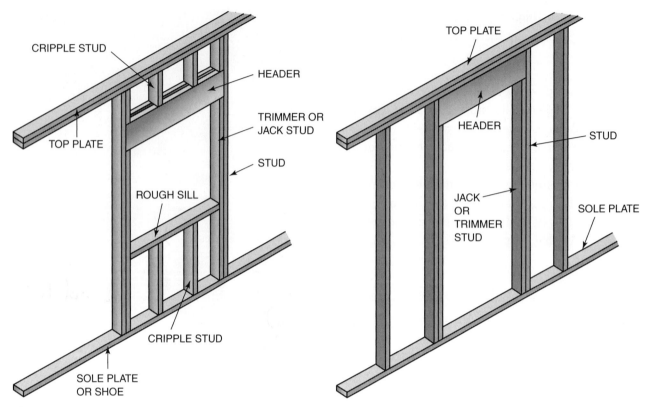

Figure 7-10: Typical framing for a window opening

Figure 7-11: Typical framing for a door opening

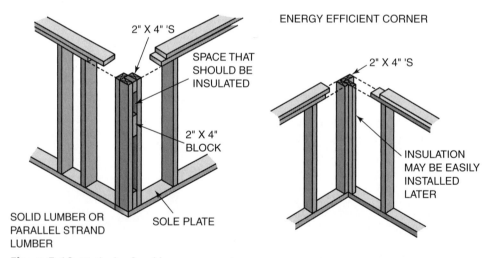

Figure 7-12: Methods of making corner posts

- Rough sills—form the bottom of a window opening at right angles to the studs (Figures 7-10 and 7-11).
- Trimmers (jacks)—shortened studs that support the headers.
- Corner Posts—same length as studs (Figure 7-12)

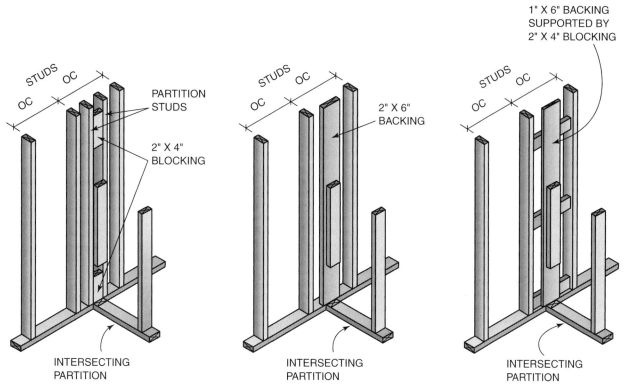

Figure 7-13: **Partition intersections are constructed in several ways**

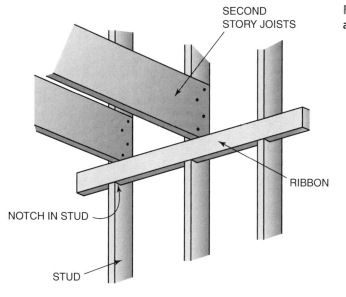

Figure 7-14: **Ribbons are used to support floor joists in a balloon frame**

- Partition intersections—framing needed when interior partitions meet an exterior partition (Figure 7-13).
- Ribbons—horizontal members of the exterior wall frame in balloon construction (Figure 7-14).
- Corner braces—used to brace walls (Figures 7-15 and 7-16).

134 RESIDENTIAL CONSTRUCTION ACADEMY FACILITIES MAINTENANCE

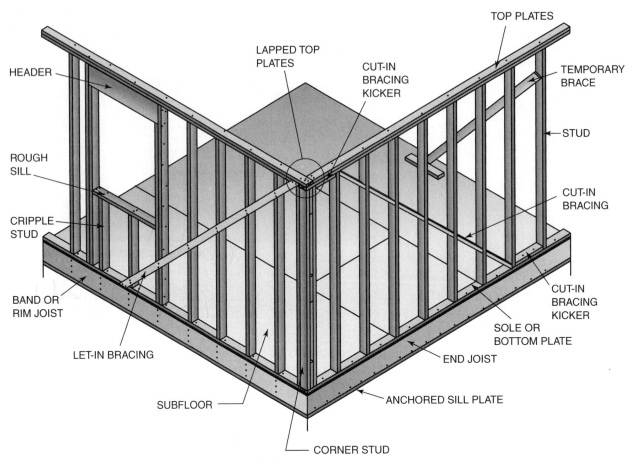

Figure 7-15: Wood wall bracing may be cut in or let in

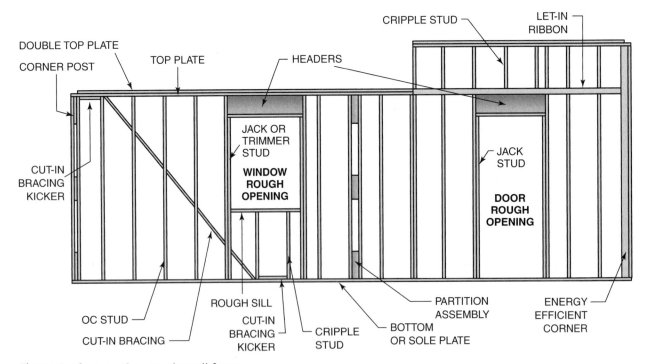

Figure 7-16: Parts of an exterior wall frame

Interior Carpentry Maintenance

Procedures

Constructing the Grid Ceiling System

 Locate the height of the ceiling, marking elevations of the ceiling at the ends of all wall sections. Snap chalk lines on all walls around the room to the height of the top edge of the wall angle. If a laser is used, the chalk line is not needed since the ceiling is built to the light beam.

- Fasten wall angles around the room with their top edge lined up with the line. Fasten into framing wherever possible, not more than 24 inches apart. If available, power nailers can be used for efficient fastening.

FROM EXPERIENCE

To fasten wall angles to concrete walls, short masonry nails sometimes are used. However, they are difficult to hold and drive. Use a small strip of cardboard to hold the nail while driving it with the hammer.

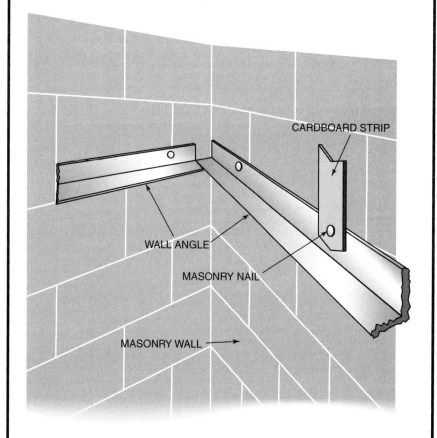

Procedures

Constructing the Grid Ceiling System (Continued)

B Make miter joints on outside corners. Make butt joints in interior corners and between straight lengths of wall angle. Use a combination square to layout and draw the square and angled lines. Cut carefully along the lines with snips.

- From the ceiling sketch, determine the position of the first main runner. Stretch a line at this location across the room from the top edges of the wall angle. The line serves as a guide for installing *hanger lags* or *screw eyes* and *hanger wires* from which main runners are suspended.

- Install the cross tee line by measuring out from the short wall, along the stretched main runner line, a distance equal to the width of the border panel. Mark the line. Stretch the cross tee line through this mark and at right angles to the main runner line.

- Install hanger lags not more than 4 feet apart and directly over the stretched line. Hanger lags should be of the type commonly used for suspended ceilings. They must be long enough to penetrate wood joists a minimum of 1 inch to provide strong support. Hanger wires may also be attached directly around the lower chord of bar joists or trusses.

CAUTION: Use care in handling the cut ends of the metal grid system. The cut ends are sharp and may have jagged edges that can cause serious injury.

B

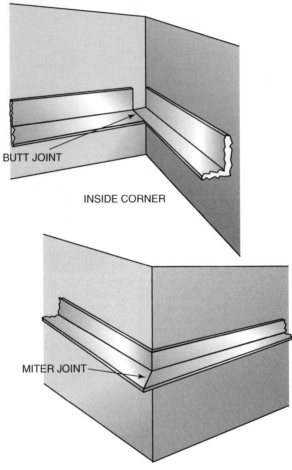

BUTT JOINT

INSIDE CORNER

MITER JOINT

OUTSIDE CORNER

FROM EXPERIENCE

Stretch the line tightly on nails inserted between the wall and wall angle.

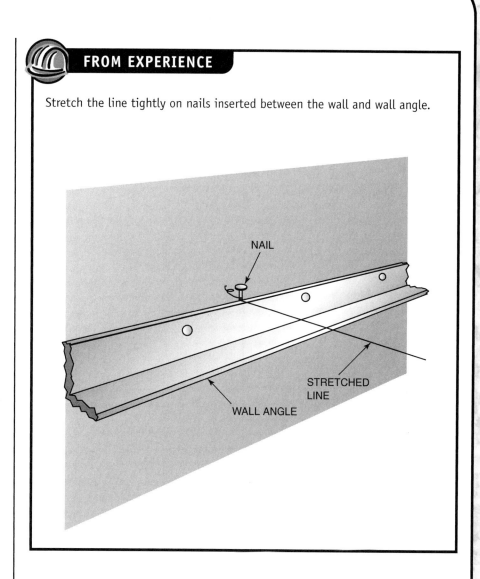

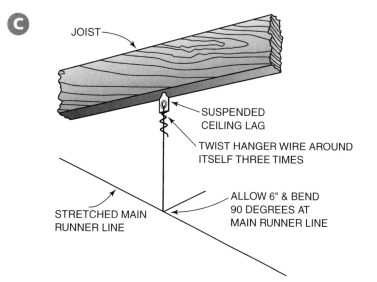

C Cut a number of hanger wires using wire cutters. The wires should be about 12 inches longer than the distance between the overhead construction and the stretched line. Attach the hanger wires to the hanger lags. Insert about 6 inches of the wire through the screw eye. Securely wrap the wire around itself three times. Pull on each wire to remove any kinks. Then make a 90-degree bend where it crosses the stretched line. If a laser is used, the 90-degree bend is done later when the main runner is installed.

Procedures

Constructing the Grid Ceiling System (Continued)

D Stretch lines, install hanger lags, and attach and bend hanger wires in the same manner at each main runner location. Leave the last line stretched tightly in position. It will be used to align the cross tee slots of the main runner.

- At each main runner location, measure from the wall to the cross tee line. Transfer this measurement to the main runner, measuring from the first cross tee slot beyond the measurement, so as to cut as little as possible from the end of the main runner.

- **EXAMPLE:** If the first cross tee will be located 23 inches from the wall, then the main runner will be cut.

- Cut the main runners about ⅛ inch less to allow for the thickness of the wall angle. Backcut the web slightly for easier installation at the wall. Measure and cut main runners individually. Do not use the first one as a pattern to cut the rest. Measure each from the cross tee line.

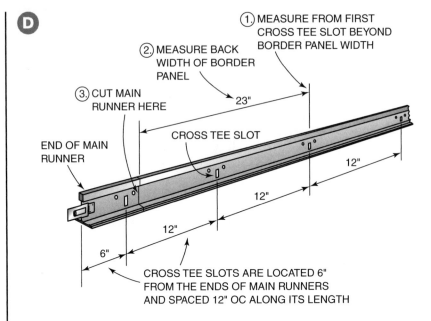

E Hang the main runners by resting the cut end on the wall angle and inserting suspension wires in the appropriate holes in the top of the main runner. Bring the runners up to the bend in the wires or to the laser light beam. Twist the wires with at least three turns to hold the main runners securely. More than one length of main runner may be needed to reach the opposite wall. Connect lengths of main runners together by inserting tabs into matching ends. Make sure end joints come up tight.

Courtesy of Trimble.

F The length of the last section is measured from the end of the last one installed to the opposite wall, allowing about ⅛ inch less to fit.

- Cross tees are installed by inserting the tabs on the ends into the slots in the main runners. These fit into position easily, although the method of attaching varies from one manufacturer to another. Install all full-length cross tees between main runners first.

- Lay in a few full-size ceiling panels to stabilize the grid while installing the border cross tees.

- Cut and install cross tees along the border. Insert the connecting tab of one end in the main runner and rest the cut end on the wall angle. It may be necessary to measure and cut cross tees for border panels individually, if walls are not straight or square.

- For 2 × 2 panels, install 2-foot cross tees at the midpoints of the 4-foot cross tees. After the grid is complete, straighten and adjust the grid to level and straight where necessary.

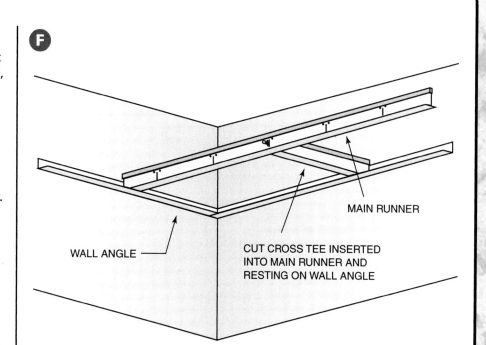

Procedures

Constructing the Grid Ceiling System (Continued)

G Ceiling panels are placed in position by tilting them slightly, lifting them above the grid, and letting them fall into place. Be careful when handling panels to avoid marring the finished surface. Cut and install border panels first and install the full-sized panels last. Measure each border panel individually, if necessary. Cut them ⅛ inch smaller than measured so they can drop into place easily. Cut the panels with a sharp utility knife using a straightedge as a guide. A scrap piece of cross tee material can be used as a straightedge. Always cut with the finished side of the panel up.

G

H When a column is near the center of a ceiling panel, cut the panel at the midpoint of the column. Cut semicircles from the cut edge to the size required for the panel pieces to fit snugly around the column. After the two pieces are rejoined around the column, glue scrap pieces of panel material to the back of the installed panel. If the column is close to the edge or end of a panel, cut the panel from the nearest edge or end to fit around the column. The small piece is also fitted around the column and joined to the panel by gluing scrap pieces to its back side.

H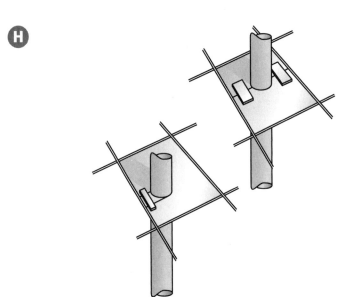

Courtesy of Armstrong World Industries.

Procedures | Applying Wall Molding

A To snap a line for wall trim, begin by holding a short scrap piece of the molding at the proper angle on the wall. Lightly mark the wall along the bottom edge of the molding. Measure the distance from the ceiling down to the mark.

- Measure and mark this same distance down from the ceiling on each end of each wall to which the molding is to be applied. Snap lines between the marks. Apply the molding so its bottom edge is to the chalk line.

B Apply the molding to the first wall with square ends in both corners. If more than one piece is required to go from corner to corner, the butt joints may be squared or mitered. Position the molding in the miter box the same way each time. Mitering the molding with the same side down each time helps make fitting more accurate, faster, and easier.

A

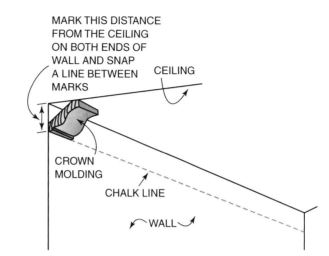

B

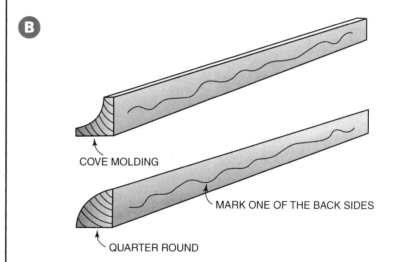

 FROM EXPERIENCE

Since the revealed edges of the molding are often not the same, it is important to cut the molding with the same orientation. To do this mark one of the back surfaces with a pencil.

Procedures: Applying Wall Molding (Continued)

C If a small-size molding is used, fasten it with finish nails in the center. Use nails of sufficient length to penetrate into solid wood at least one inch. If large-size molding is used, fastening is required along both edges. Nail at about 16-inch intervals and in other locations as necessary to bring the molding tight against the surface. End nails should be placed 2 to 3 inches from the end to keep from splitting the molding. If it is likely that the molding may split, blunt the pointed end of the nail.

- Cope the starting end of the first piece on each succeeding wall against the face of the last piece installed on the previous wall. Work around the room in one direction. The end of the last piece installed must be coped to fit against the face of the first piece.

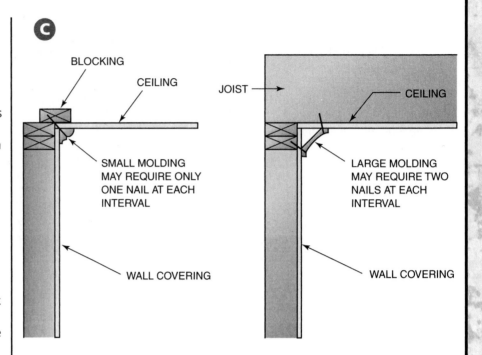

Procedures: Applying Door Casings

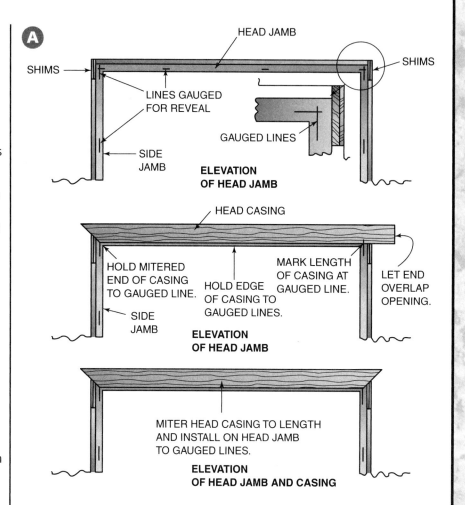

A Set the blade of the combination square so that it extends 5/16 inch beyond the body of the square. Gauge lines at intervals along the side and head jamb edges by riding the square against the inside face of the jamb. Let the lines intersect where side and head jambs meet.

- Cut one miter on the ends of the two side casings. Cut them a little long as they will be cut to fit later. Be sure to cut pairs of right and left miters.

- Miter one end of the head casing. Hold it against the head jamb of the door frame so that the miter is on the intersection of the gauged lines. Mark the length of the head casing at the intersection of the gauged lines on the opposite side of the door frame. Miter the casing to length at the mark.

- Fasten the head casing in position with a few tack nails. It may be necessary to move the ends slightly to fit the mitered joint between head and side casings. Keep the casing inside edge aligned to the gauged lines on the head jamb. The mitered ends should be in line with the gauged lines on the side jambs. Use finish nails along the inside edge of the casing into the header jamb. Straighten the casing as necessary as nailing progresses. Drive nails at the proper angle to keep them from coming through the face or back side of the jamb. Fasten the top edge of the casing into the framing.

Procedures: Applying Door Casings (Continued)

- Cut the previously mitered side casing to length. Mark the bottom end by turning it upside down with the point of the miter touching the floor. Mark the side casing in line with the top edge of the head casing. Make a square cut on the casing at that mark. If the finish floor has not been laid, hold the point of the miter on a scrap block of material that is equal in thickness to the finish floor. Replace the side casing in position and try the fit at the mitered joint. If the joint needs adjusting, trim with a power miter box or use a sharp block plane.

- When fitted, a little glue may be applied to the joint. Nail the side casing in the same manner as the head casing. Bring the faces flush, if necessary, by shimming between the back of the casing and the wall. Usually, only thin shims are needed. Any small space between the casing and the wall is usually not noticeable or it can be filled later with joint filling compound. Also, the backside of the thicker piece may be planed or chiseled to the desired thinness.

- Drive a 4d finish nail into the edge of the casing and through the mitered joint. Then set all fasteners. Keep nails 2 or 3 inches from the end to avoid splitting the casing.

FROM EXPERIENCE

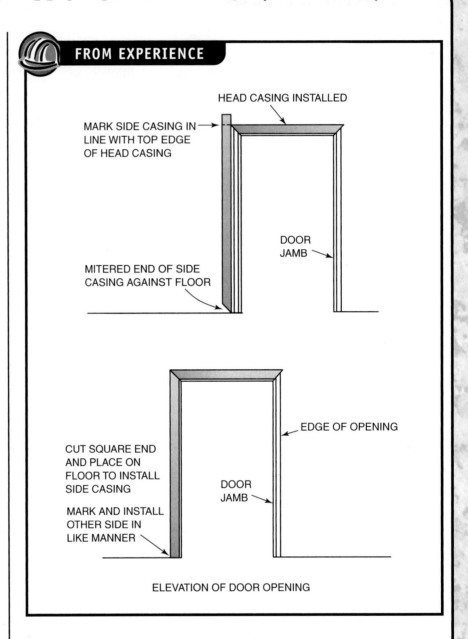

ELEVATION OF DOOR OPENING

Procedures: Hanging Interior Doors

Setting a Prehung Door Frame

 Remove the protective packing from the unit. Leave the small fiber shims between the door and jambs to help maintain this space. Cut off the horns if necessary. Remove nail that holds the door closed.

- Center the unit in the opening, so the door will swing in the desired direction. Be sure the door is closed and spacer shims are still in place between the jamb and door.

 Level the head jamb. Make adjustments by shimming the jamb that is low so it brings the head jamb level. Adjust a scriber to the thickness of shim and scribe this amount off of the other jamb. Remove frame and cut the jamb. Note the clearance under the door is being reduced by the amount being cut off.

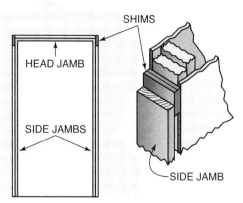

ELEVATION

SET FRAME IN OPENING. SHIM ON BOTH SIDES OPPOSITE HEAD JAMB. LEVEL HEAD JAMB AND FASTEN AT TOP.

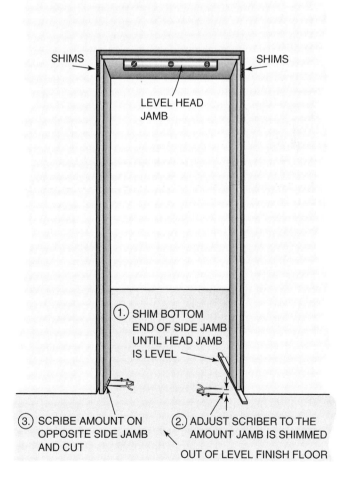

Procedures: Hanging Interior Doors (Continued)

C Plumb the hinge side jamb of the door unit. A two-foot carpenter's level may not be accurate when plumbing the sides because of any bow that may be in the jambs. Use a 6-foot level or a plumb bob. Tack the jamb plumb to the wall through the casing with one nail on either side.

D Open the door and move to the other side. Check that the unit is nearly centered. Install shims between the side jambs and the rough opening at intermediate points, keeping side jambs straight. Shims should be located behind the hinges and lockset **strike plates.** Nail through the side jambs and shims. Remove spacers from door edges.

- Check the operation of the door. Make any necessary adjustments. The space between the door and the jamb should be equal on all jambs. The entire door edge should touch the stop or weatherstripping.

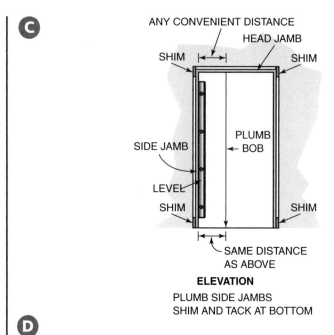

C ELEVATION
PLUMB SIDE JAMBS
SHIM AND TACK AT BOTTOM

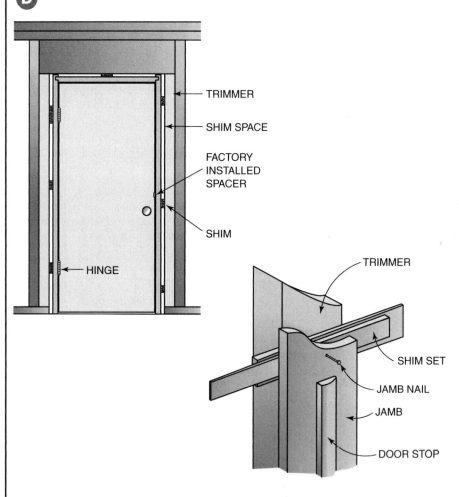

D

E Finish nailing the casing and install casing on the other side of the door. Drive and set all nails. Do not make any hammer marks on the finish.

E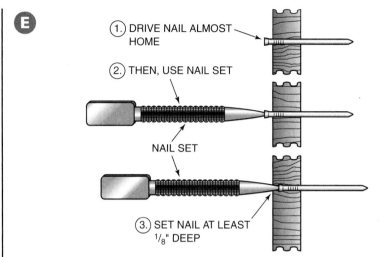

1. DRIVE NAIL ALMOST HOME
2. THEN, USE NAIL SET — NAIL SET
3. SET NAIL AT LEAST 1/8" DEEP

Fitting a Door to a Frame

A Begin by checking the door for its beveled edge and the direction of the face of the door. Note the direction of the swing.

- Lightly mark the location of the hinges on the door. On paneled doors, the top hinge is usually placed with its upper end in line with the bottom edge of the top rail. The bottom hinge is placed with its lower end in line with the top edge of the bottom rail. The middle hinge is centered between them. On flush doors, the usual placement of the hinge is approximately 9 inches down from the top and 13 inches up from the bottom, as measured to the center of the hinge. A middle hinge is centered between the two.

- Check the opening frame for level and plumb.

A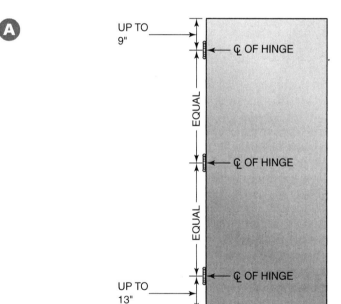

UP TO 9" — ℄ OF HINGE
EQUAL
℄ OF HINGE
EQUAL
UP TO 13" — ℄ OF HINGE

Procedures: Hanging Interior Doors (Continued)

B Plane the door edges so the door fits onto the opening with an even joint of approximately 3/32 inch between the door and the frame on all sides. A wider joint of approximately 1/8 inch must be made to allow for swelling of the door and frame in extremely damp weather. Use a *door jack* to hold the door steady. Do not cut more than 1/2 inch total from the width of a door. Cut no more than 2 inches from its height. Check the fit frequently by placing the door in the opening, even if this takes a little extra time.

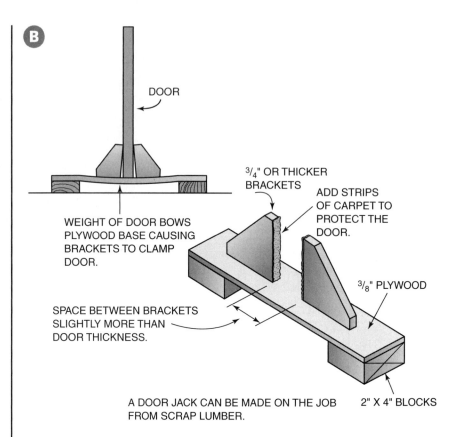

B A DOOR JACK CAN BE MADE ON THE JOB FROM SCRAP LUMBER.

C Place the door in the frame. Shim the door so the proper joint is obtained along all sides. Place shims between the lock edge of the door and side jamb of the frame. Mark across the door and jamb at the desired location for each hinge. Place a small X on both the door and the jamb, to indicate on which side of the mark to cut the gain.

- Remove the door from the frame. Place a hinge leaf on the door edge with its end on the mark previously made. Score a line along edges of the leaf. Score only partway across the door edge.

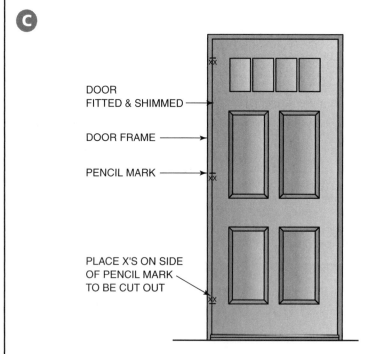

D Score the hinge lines, taking care not to split any part of the door. With a chisel, cut small chips from each end of the gain joint. The chips will break off at the scored end marks. Then, with the flat of the chisel down, pare and smooth the excess down to the depth of the gain. Be careful not to slip.

- Press the hinge leaf into the gain joint. It should be flush with the door edge and install screws. Center the screws carefully so the hinge leaf will not shift when the screw head comes in contact with the leaf.

- Place the door in the opening and insert the hinge pins. Check the swing of the door and adjust as needed.

E Apply the *door stops* to the jambs with several tack nails, in case they have to be adjusted when locksets are installed. A **back miter** joint is usually made between molded side and header stops. A butt joint is made between square-edge stops.

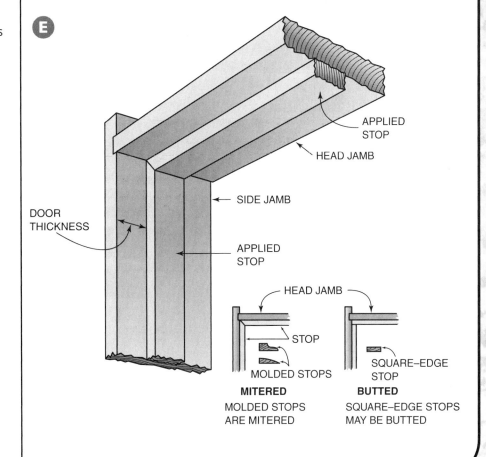

Procedures: Hanging Interior Doors (Continued)

Installing Bypass Doors

A Cut the track to length. Install it on the header jamb according to the manufacturer's directions. Bypass doors are installed so they overlap each other by about 1 inch when closed.

- Install pairs of *roller hangers* on each door. The roller hangers may be offset a different amount for the door on the outside than the door on the inside. They are also offset differently for doors of various thicknesses. Make sure that rollers with the same and correct offset are used on each door. The location of the rollers from the edge of the door is usually specified in the manufacturer's instruction sheet.

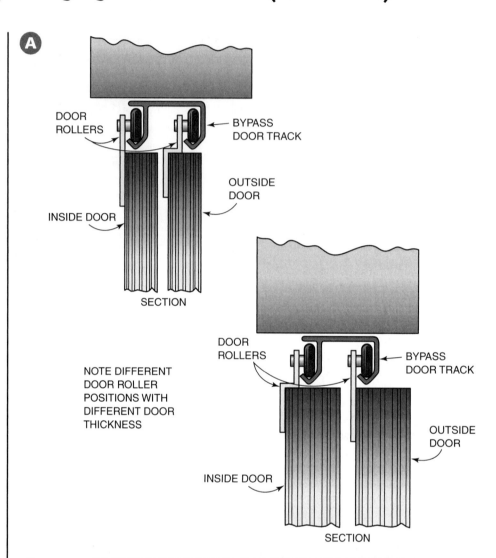

B Mark the location and bore holes for *door pulls*. Flush pulls must be used so that bypassing is not obstructed. The proper size hole is bored partway into the door. The pull is tapped into place with a hammer and wood block. The press fit holds the pull in place. Rectangular flush pulls, also used on bypass doors, are held in place with small recessed screws.

C Hang the doors by holding the bottom outward. Insert the rollers in the overhead track. Then gently let the door come to a vertical position. Install the inside door first, then the outside door.

- Test the door operation and the fit against side jambs. Door edges must fit against side jambs evenly from top to bottom. If the top or bottom portion of the edge strikes the side jamb first, it may cause the door to jump from the track. The door rollers have adjustments for raising and lowering. Adjust one or the other to make the door edges fit against side jambs.

D A *floor guide* is included with bypass door hardware to keep the doors in alignment. The guide is centered on the lap of the two doors to steady them at the bottom. Mark the location and fasten the guide.

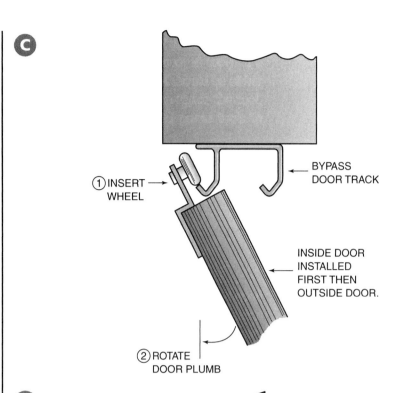

C
① INSERT WHEEL
BYPASS DOOR TRACK
INSIDE DOOR INSTALLED FIRST THEN OUTSIDE DOOR.
② ROTATE DOOR PLUMB

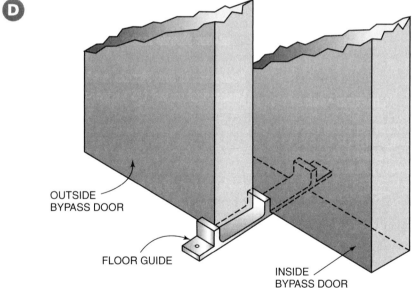

D
OUTSIDE BYPASS DOOR
FLOOR GUIDE
INSIDE BYPASS DOOR
FLOOR GUIDE IS ADJUSTABLE FOR VARIOUS DOOR THICKNESSES

Procedures

Hanging Interior Doors (Continued)

Installing Bifold Doors

A Check that the door and its hardware are all present. The hardware consists of the track, pivot sockets, pivot pins and guides, door aligners, door pulls, and necessary fasteners.

B Cut the track to length. Fasten it to the header jamb with screws provided in the kit. The track contains adjustable *sockets* for the door *pivot pins*. Make sure these are inserted before fastening the track in position. The position of the track on the header jamb is not critical. It may be positioned as desired.

- Locate the bottom pivot sockets. Fasten one on each side, at the bottom of the opening. The pivot socket bracket is L-shaped. It rests on the floor against the side jamb. It is centered on a plumb line from the center of the pivot sockets in the track on the header jamb above.

- Install pivot pins at the top and bottom ends of the door in the prebored holes closest to the jamb. Sometimes the top pivot pin is spring loaded. It can then be depressed for easier installation of the door. The bottom pivot pin is threaded and can be adjusted for height. The guide pin rides in the track. It is installed in the hole provided at the top end of the door farthest away from the jamb.

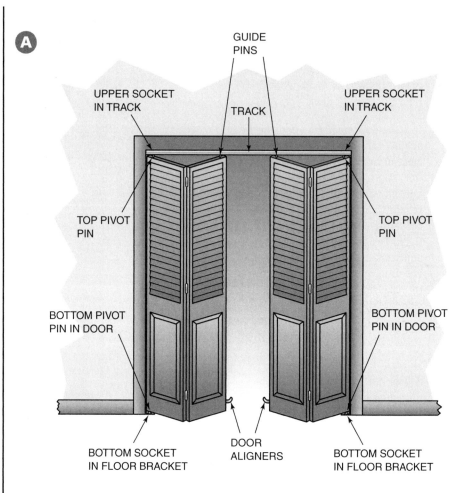

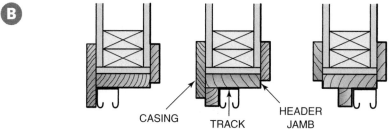

Loosen the set screw in the top pivot socket. Slide the socket along the track toward the center of the opening about one foot away from the side jamb. Place the bottom door pivot in position by inserting it into the bottom pivot socket. Tilt the door to an upright position, while at the same time inserting the top pivot pin in the top socket. Slide the top pivot socket back toward the jamb where it started from.

- Adjust top and bottom pivot sockets in or out so the desired joint is obtained between the door and the jamb. Lock top and bottom pivot sockets in position. Adjust the bottom pivot pin to raise or lower the doors, if necessary.

- Install second door in the same manner.

- Install pull knobs and door aligners in the manner and location recommended by the manufacturer. The door aligners keep the faces of the center doors lined up when closed.

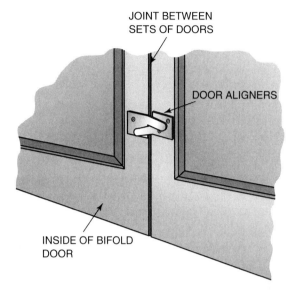

Procedures: Applying Base Moldings

A Cut the first piece with squared ends if it fits between two walls. Miter the butt joint if desired. If one piece fits from corner to corner, its length may be determined by measuring from corner to corner. Then, transfer the measurement to the baseboard. Cut the piece ½ to 1 inch longer. Place the piece in position with one end tight to the corner and the other end away from the corner. Press the piece tight to the wall near the center. Place small marks on the top of the base trim and onto the wall so they line up with each other. Reposition the piece with the other end in the corner. Press the base against the wall at the mark. The difference between the mark on the wall and the mark on the base is the amount to cut off.

- After cutting, place one end of the piece in the corner and bow out the center. Place the other end in the opposite corner, and press the center against the wall. Fasten in place. Continue in this manner around the room. Make regular miter joints on outside corners.

- If both ends of a single piece are to have regular miters for outside corners, it is important that it be fastened in the same position as it was marked. Tack the rough length in position with one finish nail in the center. Mark both ends. Remove, and cut the miters. Remember that these marks are to the short side of the miter so the piece will be longer than these marks indicate. Reinstall the piece by first fastening into the original nail hole.

FROM EXPERIENCE

Use this method of cutting a full length molding to fit between corners.

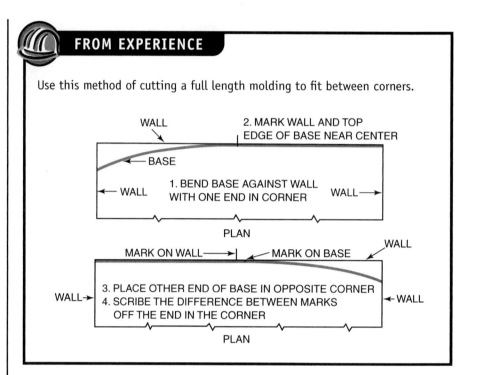

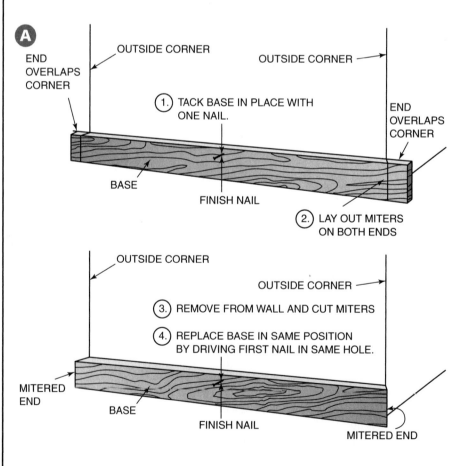

B If a *base cap* is applied it is done so in the same manner as most wall or ceiling molding. Cope interior corners and miter exterior corners. However, it should be nailed into the floor and not into the baseboard. This prevents the joint under the shoe from opening should shrinkage take place in the baseboard.

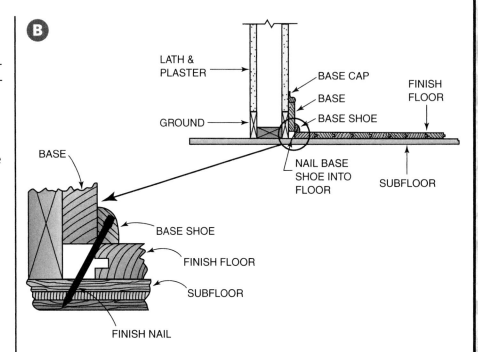

C When the base shoe must be stopped at a door opening or other location, with nothing to butt against, its exposed end is generally *back-mitered* and sanded smooth. Generally, no base shoe is required if carpeting is to be used as a floor finish.

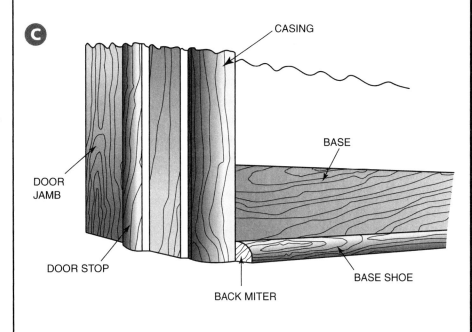

Procedures: Installing Window Trim

Applying the Stool

A Hold a piece of side casing in position at the bottom of the window and draw a light line on the wall along the outside edge of the casing stock. Mark a distance outward from these lines equal to the thickness of the window casing. Cut a piece of stool stock to length equal to the distance between the outermost marks.

- Position the stool with its outside edge against the wall. The ends should be in line with the marks previously made on the wall. Lightly square lines, across the face of the stool, even with the inside face of each side jamb of the window frame.

- Set the pencil dividers or scribers to mark the cutout so that, on both sides, an amount equal to twice the casing thickness will be left on the stool. Scribe the stool by riding the dividers along the wall on both sides. Also scribe along the bottom rail of the window sash.

- Cut to the lines, using a handsaw. Smooth the sawed edge that will be nearest to the sash. Shape and smooth both ends of the stool the same as the inside edge.

- Apply a small amount of caulking compound to the bottom of the stool. Fasten the stool in position by driving finish nails along its outside edge into the sill. Set the nails.

> **FROM EXPERIENCE**
>
> Raise the lower sash slightly. Place a short, thin strip of wood under it, on each side, which projects inward to support the stool while it is being laid out. Place the stool on the strips. Raise or lower the sash slightly so the top of the stool is level.

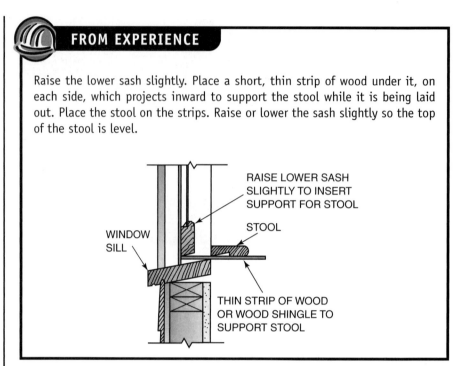

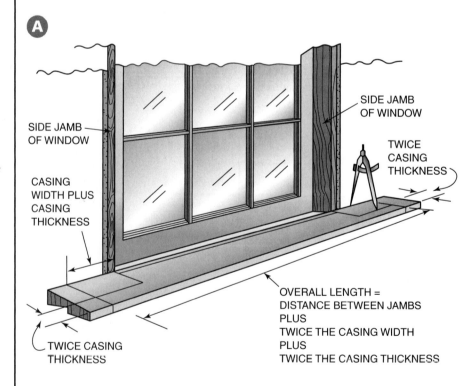

Applying the Apron

 Cut a length of apron stock equal to the distance between the outer edges of the window casings.

- Each end of the apron is then *returned upon itself*. This means that the ends are shaped the same as its face. To return an end upon itself, hold a scrap piece on the apron. Draw its profile flush with the end. Cut to the line with a coping saw. Sand the cut end smooth. Return the other end upon itself in the same manner.

- Place the apron in position with its upper edge against the bottom of the stool. Be careful not to force the stool upward. Keep the top side of the stool level by holding a square between it and the edge of the side jamb. Fasten the apron along its bottom edge into the wall. Then drive nails through the stool into the top edge of the apron.

FROM EXPERIENCE

When nailing through the stool, wedge a short length of 1 × 4 stock between the apron and the floor at each nail location. This supports the apron while nails are being driven. Failure to support the apron results in an open joint between it and the stool. Take care not to damage the bottom edge of the apron with the supporting piece.

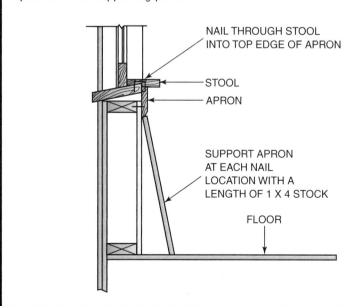

Procedures: Installing Window Trim (Continued)

Installing Jamb Extensions

A Measure the distance from the jamb to the finished wall. Rip the jamb extensions to this width with a slight back-bevel on the side toward the jack stud.

- Cut the pieces to length and apply them to the header, side jambs, and stool. Shim them, if necessary, and nail with finish nails that will penetrate the framing at least an inch.

Applying the Casings

- Cut the number of window casings needed to a rough length with a miter on one end. Cut side casings with left- and right-hand miters.

- Install the header casing first and then the side casings in a similar manner as with door casing. Find the length of side casings by turning them upside down with the point of the miter on the stool in the same manner as door casings.

- Fasten casings with their inside edges flush with the inside face of the jamb or with a reveal. Make neat, tight-fitting joints at the stool and at the head.

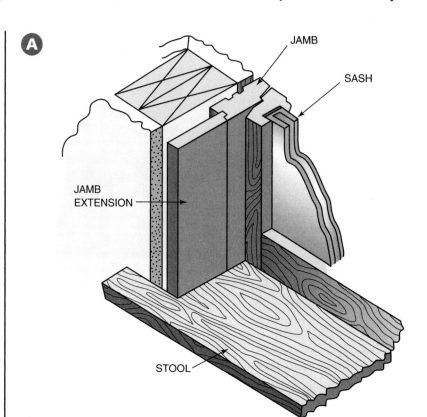

Procedures

Installing Wood Flooring

Preparation for Installation

- Check the subfloor for any loose areas and add nails where appropriate. Sweep and vacuum the subfloor clean. Scraping may be necessary to remove any unwanted material.

- Cover the subfloor with building paper. Lap it 4 inches at the seams, and at right angles to the direction of the finish floor. The paper prevents squeaks in dry seasons and retards moisture from below that could cause warping of the floor.

- Snap chalk lines on the paper showing the centerline of floor joists so flooring can be nailed into them. For better holding power, fasten flooring through the subfloor and into the floor joists whenever possible. On ½-inch plywood subfloors, all flooring fasteners must penetrate into the joists.

Starting Strip

- The location and straight alignment of the first course is important. Place a strip of flooring on each end of the room, ¾ inch from the starter wall with the groove side toward the wall. The gap between the flooring and the wall is needed for expansion. It will eventually be covered by the base molding.

- Mark along the edge of the flooring tongue. Snap a chalk line between the two points. Hold the strip with its tongue edge to the chalk line.

- **A** Face nail it with 8d finish nails, alternating from one edge to the other, 12 to 16 inches apart.

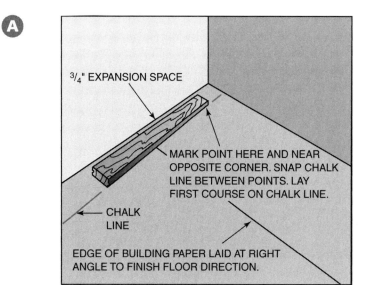

Courtesy of Chickasaw Hardwood Floors.

Procedures: Installing Wood Flooring (Continued)

Make sure end joints between strips are driven up tight.

- Cut the last strip to fit loosely against the wall. Use a strip long enough so that the cut-off piece is 8 inches or longer. This scrap piece is used to start the next course back against the other wall.

B After the second course of flooring is fastened, lay out seven or eight loose rows of flooring, end to end. This is called *racking the floor*. Racking is done to save time and material.

- Lay out in a staggered end-joint pattern. End joints should be at least 6 inches apart. Find or cut pieces to fit within ½ inch of the end wall. Distribute long and short pieces evenly for the best appearance. Avoid clusters of short strips. Lay out loose flooring.

- Continue across the room. Rack seven or eight courses ahead as work progresses.

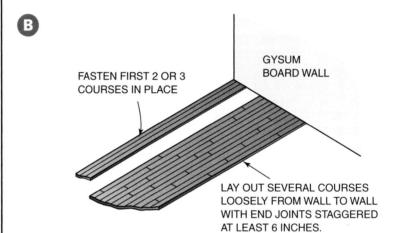

B

FASTEN FIRST 2 OR 3 COURSES IN PLACE

GYSUM BOARD WALL

LAY OUT SEVERAL COURSES LOOSELY FROM WALL TO WALL WITH END JOINTS STAGGERED AT LEAST 6 INCHES.

CHAPTER 7 Carpentry

Procedures: Installing Manufactured Cabinets

Cabinet Layout Lines

A Measure 34½ inches up the wall. Draw a level line to indicate the tops of the base cabinets. Measure and mark another level line on the wall 54 inches from the floor. The bottom of the wall units are installed to this line.

- Next mark the stud locations of the framed wall. Cabinet mounting screws will be driven into the studs. Lightly tap on and across a short distance of the wall with a hammer. Above the upper line on the wall, drive a finish nail in at the point where a solid sound is heard to accurately locate the stud. Drive nails where the holes will be later covered by a cabinet. Mark the locations of the remaining studs where cabinets will be attached. At each stud location, draw plumb lines on the wall. Mark the outlines of all cabinets on the wall to visualize and check the cabinet locations against the layout.

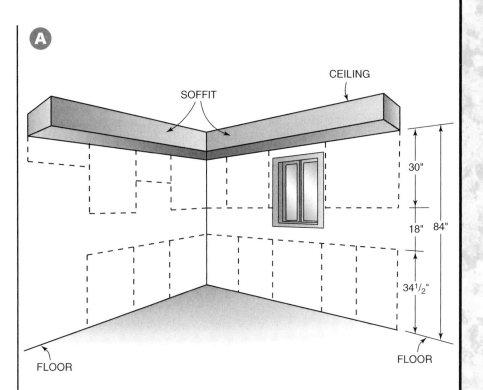

Installing Wall Units

A A *cabinet lift* may be used to hold the cabinets in position for fastening to the wall. If a lift is not available, the doors and shelves may be removed to make the cabinet lighter and easier to clamp together. If possible, screw a strip of lumber so its top edge is on the level line for the bottom of the wall cabinets or strips of wood cut to the proper length. This is used to support the wall units while they are being fastened. If it is not possible to screw to the wall, build a stand on which to support the unit near the line of installation.

Procedures

Installing Manufactured Cabinets (Continued)

B Start the installation of wall cabinets in a corner. On the wall, measure from the line representing the outside of the cabinet to the stud centers. Transfer the measurements to the cabinets. Drill shank holes for mounting screws through mounting rails usually installed at the top and bottom of the cabinet. Place the cabinet on the supporting strip or stand so its bottom is on the level layout line. Fasten the cabinet in place with mounting screws of sufficient length to hold the cabinet securely. Do not fully tighten the screws. The next cabinet is installed in the same manner.

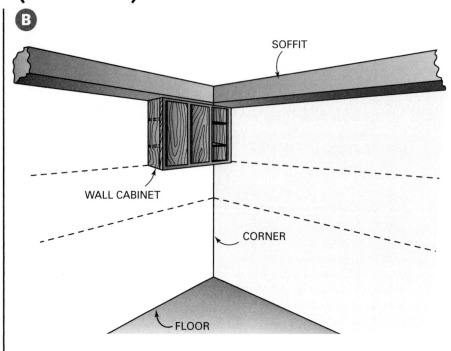

C Align the adjoining *stiles* so their faces are flush with each other. Clamp them together with C-clamps. Screw the stiles tightly together. Continue this procedure around the room. After all the stiles are secured to each other, tighten all mounting screws. If a filler needs to be used, it is better to add it at the end of a run. It may be necessary to scribe the filler to the wall.

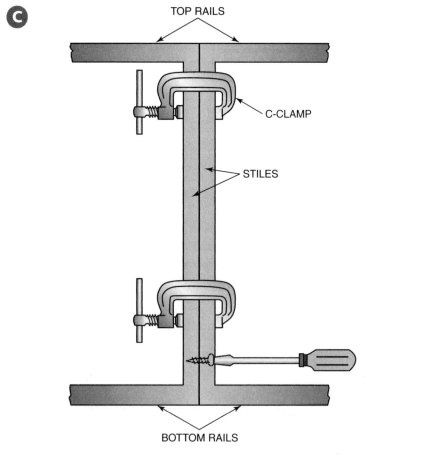

- Procedure for scribing a filler strip at the end of a run of cabinets.
- The space between the top of the wall unit and the ceiling may be finished by installing a soffit.

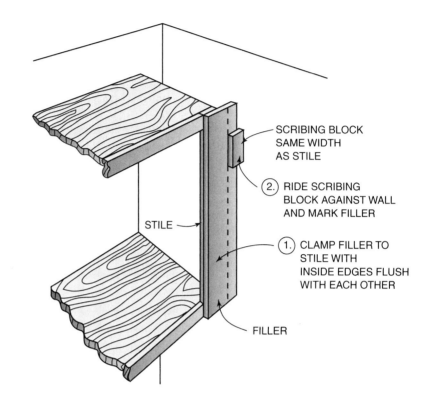

Installing Base Cabinets

A Start the installation of base cabinets in a corner. Shim the bottom until the cabinet top is on the layout line. Then level and shim the cabinet from back to front. If cabinets are to be fitted to the floor, shim until their tops are level across width and depth. This will bring the tops above the layout line that was measured from the low point of the floor. Adjust the scriber so the distance between the points is equal to the amount the top of the unit is above the layout line. Scribe this amount on the bottom end of the cabinets by running the dividers along the floor.

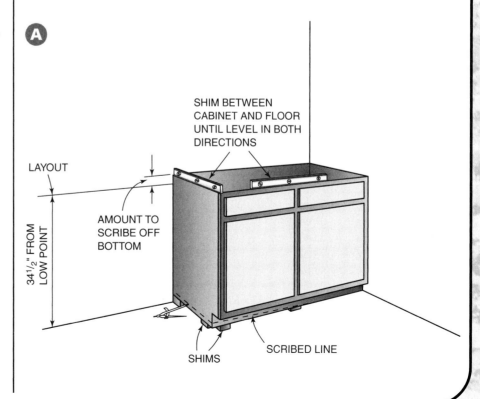

Procedures

Installing Manufactured Cabinets (Continued)

 Cut both ends and toeboard to the scribed lines. Replace the cabinet in position. The top ends should be on the layout line. Fasten it loosely to the wall. The remaining base cabinets are installed in the same manner. Align and clamp the stiles of adjoining cabinets. Fasten them together. Finally, fasten all units securely to the wall.

Installing Countertops

- After the base units are fastened in position, the countertop is cut to length. It is fastened on top of the base units and against the wall. The backsplash can be scribed, limited by the thickness of its scribing strip, to an irregular wall surface. Use pencil dividers to scribe a line on the top edge of the backsplash. Then plane or belt sand to the scribed line.

- Fasten the countertop to the base cabinets with screws up through triangular blocks usually installed in the top corners of base units. Take care not to drill through the countertop. Use screws of sufficient length, but not so long that they penetrate the countertop.

- Exposed cut ends of postformed countertops are covered by specially shaped pieces of plastic laminate. Sink cutouts are made by carefully outlining the cutout and cutting with a saber saw or router. The cutout pattern usually comes with the sink. Use a fine tooth blade to prevent chipping out the face of the laminate beyond the sink. Some duct tape applied to the base of the power tool will prevent scratching of the countertop when making the cutout.

Installing Cylindrical Locksets

To install a cylindrical lockset, first check the contents and read the manufacturer's directions carefully. There are so many kinds of locks manufactured that the mechanisms vary greatly. The directions included with the lockset must be followed carefully.

However, there are certain basic procedures. Open the door to a convenient position. Wedge the bottom to hold it in place (Figure 7-17). Measure up, from the floor, the recommended distance to the centerline of the lock. This is usually 36 to 40 inches. At this height, square a light line across the edge and stile of the door.

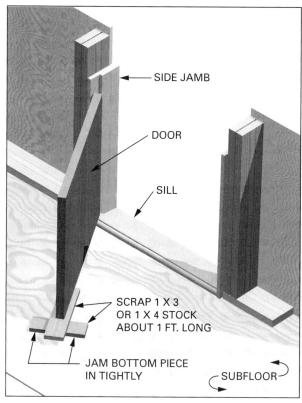

Figure 7-17: A door may be shimmed from the floor to hold it plumb during installation

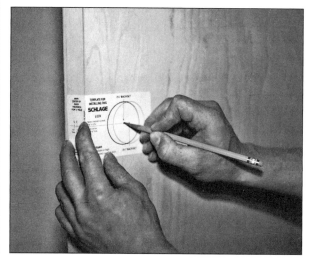

Figure 7-18: Using a template to lay out the centers of the holes for a lockset

Marking and Boring Holes

Position the center of the paper template supplied with the lock on the squared lines. Lay out the centers of the holes that need to be bored (Figure 7-18). It is important that the template be folded over the high corner of the beveled door edge. The distance from the door edge to the center of the hole through the side of the door is called the *backset* of the lock. Usual backsets are 2⅜ inches for residential and 2¾ inches for commercial. Make sure the backset is marked correctly before boring the hole. One hole must be bored through the side and one into the edge of the door.

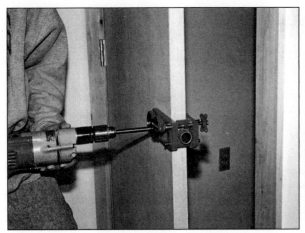

Figure 7-19: Boring jigs are frequently used to guide bits when boring holes for locksets

Figure 7-20: Using a faceplate marker

The manufacturers' directions specify the hole sizes where a 1-inch hole for bolts and 2 1/8-inch hole for locksets are common.

The hole through the side of the door should be bored first. Stock for the center of the boring bit is lost if the hole in the edge of the door is bored first. It can be bored with hand tools, using an expansion bit in a bit brace. However, it is a difficult job. If using hand tools, bore from one side until only the point of the bit comes through. Then bore from the other side to avoid splintering the door.

Using a Boring Jig. A **boring jig** is frequently used. It is clamped to the door to guide power-driven **multispur bits.** With a boring jig, holes can be bored completely through the door from one side. The clamping action of the jig prevents splintering (Figure 7-19).

After the holes are bored, insert the latchbolt in the hole bored in the door edge. Hold it firmly and score around its faceplate with a sharp knife. Remove the latch unit. Deepen the vertical lines with the knife in the same manner as with hinges. Take great care when using a chisel along these lines. This may split out the edge of the door. Then, chisel out the recess so that the faceplate of the latch lays flush with the door edge.

Faceplate markers, if available, may be used to lay out the mortise for the latch faceplate. A marker of the appropriate size is held in the bored latch hole and tapped with a hammer (Figure 7-20). Complete the installation of the lockset by following specific manufacturers' directions.

Installing the Striker Plate

The **striker plate** is installed on the door jamb so when the door is closed it latches tightly with no play. If the plate is installed too far out, the door will not close tightly against the stop. It will then rattle. If the plate is installed too far in, the door will not latch.

To locate the striker plate in the correct position, place it over the latch in the door. Close the door snugly against the stops. Push the striker plate in against the latch. Draw a vertical line on the face of the plate flush with the outside face of the door (Figure 7-21).

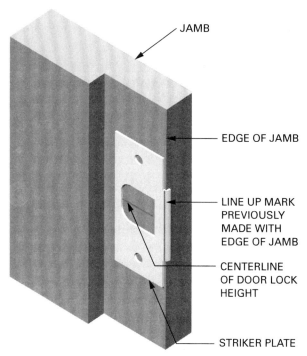

Figure 7-21: **Installing the striker plate**

Open the door. Place the striker plate on the jamb. The vertical line, previously drawn on it, should be in line with the edge of the jamb. Center the plate on the latch. Hold it firmly while scoring a line around the plate with a sharp knife. Chisel out the mortise so the plate lies flush with the jamb. Screw the plate in place. Chisel out the center to receive the latch.

Re-key Locks

1. Take the locking part of the lock out by unscrewing it from the inside. Leave the rest of the lock in place.
2. Take the knob with the locking mechanism with its proper key to the hardware store. The hardware store person will replace the pins so they'll fit another key.
3. Put the door lock back together.

Procedures: Cutting and Fitting Gypsum Board

A Take measurements accurately to within ¼ inch for the ceiling and ⅛ inch for the walls. Using a utility knife, cut the board by first scoring the face side through the paper to the core. Guide it with a *drywall T-square* using your toe to hold the bottom. Only the paper facing need be cut.

B Bend the board back against the cut. The board will break along the cut face. Score the backside paper.

FROM EXPERIENCE

Cut only the center section of the backside paper, leaving the bottom and top portions. These will act as hinges for the cut piece when it is snapped back into place.

- Lifting the panel off the floor, snap the cut piece back quickly to the straight position. This will complete the break.

C To make cuts parallel to the long edges, the board is often gauged with a tape and scored with a utility knife. When making cuts close to long edges, it is usually necessary to score both sides of the board before the break to obtain a clean cut.

- Ragged edges can be smoothed with a drywall rasp, coarse sanding block, or knife.

A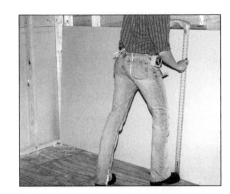

B

C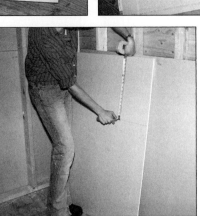

Procedures: Installing Sheet Paneling

Starting the Application

A Mark the location of each stud in the wall on the floor and ceiling. Paneling edges must fall on stud centers, even if applied with adhesive over a backer board, in case supplemental nailing of the edges is necessary.

- If the wall is to be wainscoted, snap a horizontal line across the wall to indicate its height.

- Apply narrow strips of paint on the wall from floor to ceiling over the stud where a seam in the paneling will occur. The color should be close to the color of the seams of the paneling. This will hide the joints between sheets if they open slightly because of shrinkage.

- Cut the first sheet to a length about 1/4-inch less than the wall height. Place the sheet in the corner. Plumb the edge and tack it temporarily into position.

B Notice the joint at the corner and the distance the sheet edge overlaps the stud. Set the distance between the points of a scriber to the same as the amount the sheet overlaps the center of the stud. Scribe this amount on the edge of the sheet butting the corner.

- Remove the sheet from the wall and cut close to the scribed line. Plane the edge to the line to complete the cut. Replace the sheet with the cut edge fitting snugly in the corner.

- If a tight fit between the panel and ceiling is desired, set the dividers and scribe a small amount at the ceiling line. Remove the sheet again. Cut to the scribed line. The joint at the ceiling need not be fit tight if a molding is to be used.

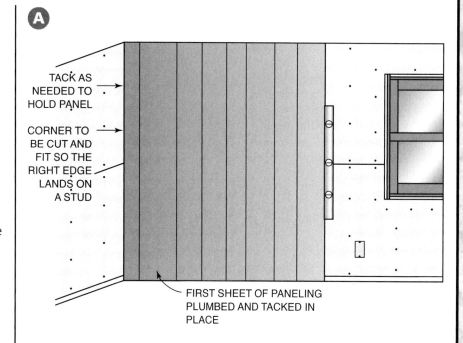

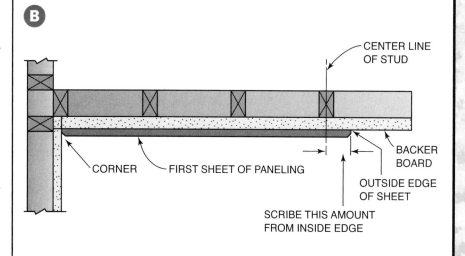

Procedures: Installing Sheet Paneling (Continued)

Wall Outlets

A To lay out for wall outlets, plumb and mark both sides of the outlet to the floor or ceiling, whichever is closer. Level the top and bottom of the outlet on the wall out beyond the edge of the sheet to be installed.

- Place the sheet in position and tack. Level and plumb marks from the wall and floor onto the sheet for the location of the opening.

B Remove the sheet. Cut the opening for the outlet. When using a saber saw, cut from the back of the panel to avoid splintering the face.

Fastening

- Apply adhesive beads 3 inches long and about 6 inches apart on all intermediate studs. Apply a continuous bead along the perimeter of the sheet. Put the panel in place. Tack it at the top when panel is in proper position.

- Press on the panel surface to make contact with the adhesive. Use firm, uniform pressure to spread the adhesive beads evenly between the wall and the panel. Then, grasp the panel and slowly pull the bottom of the sheet a few inches away from the wall.

- Press the sheet back into position after about two minutes. Drive nails as needed and recheck the sheet for a complete bond after about 20 minutes. Apply pressure to assure thorough adhesion and to smooth the panel surface.

- Apply successive sheets in the same manner. Panels should touch only lightly at joints.

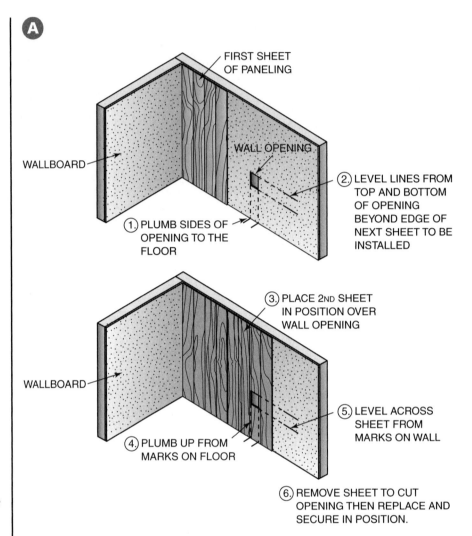

Ending the Application

- Take measurements at the top, center, and bottom. Cut the sheet to width and install. If no corner molding is used, the sheet must be cut to fit snugly in the corner. To mark the sheet accurately, first measure the remaining space at the top, bottom, and about the center. Rip the panel about ½ inch wider than the greatest distance.

A Place the sheet plumb with the cut edge in the corner and the other edge overlapping the last sheet installed. Tack the sheet in position so the amount of overlap is exactly the same from top to bottom. Set the scriber for the amount of overlap. Scribe this amount on the edge in the corner.

- Cut close to the scribed line and then plane to the line. If the line is followed carefully, the sheet should fit snugly between the last sheet installed and the corner, regardless of any irregularities.

B Exterior corners may be finished by capping the joint.

- Use a wood block for more accurate scribing of wide distances.

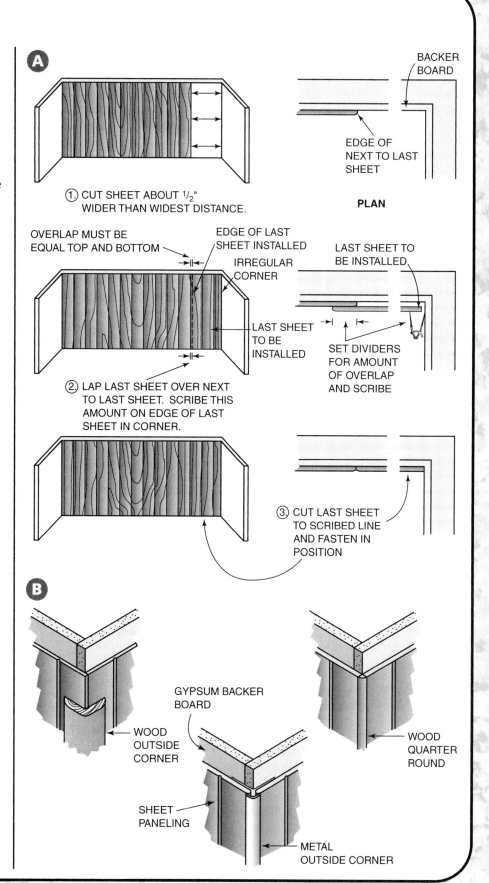

Procedures

Installing Solid Wood Paneling

Starting the Application

 Select a straight board with which to start. Cut it to length, about ¼ inch less than the height of the wall. If tongue-and-grooved stock is used, tack it in a plumb position with the grooved edge in the corner.

- Adjust the scribers to scribe an amount a little more than the depth of the groove. Rip and plane to the scribed line.

- Replace the piece and face nail along the cut edge into the corner with finish nails about 16 inches apart. Blind nail the other edge through the tongue.

- Apply succeeding boards by blind nailing into the tongue only. Make sure the joints between boards come up tightly. Severely warped boards should not be used.

- As installation progresses, check the paneling for plumb. If out of plumb, gradually bring back by driving one end of several boards a little tighter than the other end. Cut out openings in the same manner as described for sheet paneling.

Applying the Last Board

 Cut and fit the next to the last board. Then remove it. Cut, fit and tack the last board in the place of the next-to-the-last board just previously removed.

- Cut a scrap block about 6 inches long and equal in width to the finished face of the next-to-the-last board. The tongue should be removed. Use this block to scribe the last board by running one edge along the corner and holding a pencil against the other edge.

- Remove the board from the wall. Cut and plane it to the scribed line. Fasten the next-to-the-last board in position. Fasten the last board in position with the cut edge in the corner.

- Face nail the edge nearest the corner.

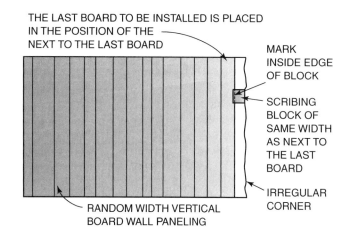

Procedures: Installing Flexible Insulation

 Install positive ventilation chutes between the rafters where they meet the wall plate. This will compress the insulation slightly against the top of the wall plate to permit the free flow of air over the top of the insulation.

- Install the air-insulation dam between rafters in line or on with the exterior sheathing. This will protect the insulation from air movement into the insulation layer from the soffits.

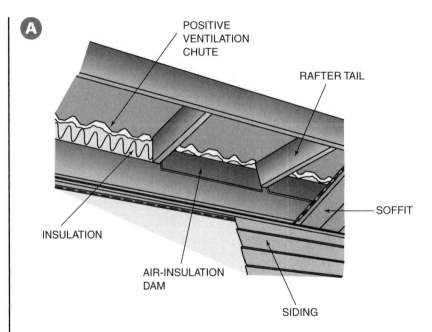

 To cut the material, place a scrap piece of plywood on the floor to protect the floor while cutting. Roll out the material over the scrap. Using another scrap piece of wood, compress the insulation and cut it with a sharp knife in one pass.

Place the batts or blankets between the studs. The flanges of the vapor retarder may be stapled either to cover the studs or to the inside edges of the studs as well as the top and bottom plates. A better vapor retarder is achieved with fastening to cover the stud, but the studs are less visible for the installation of the gypsum. Use a hand or hammer-tacker stapler to fasten the insulation in place.

D Fill any spaces around windows and doors with spray-can foam. Non-expanding foam will fill the voids with an airtight seal and protect the house from air leakage. After the foam cures, flexible insulation may be added to fill the remaining space.

- Install ceiling insulation by stapling it to the ceiling joists or by friction-fitting it between them. Push and extend the insulation across the top plate to fit against the air-insulation dam.

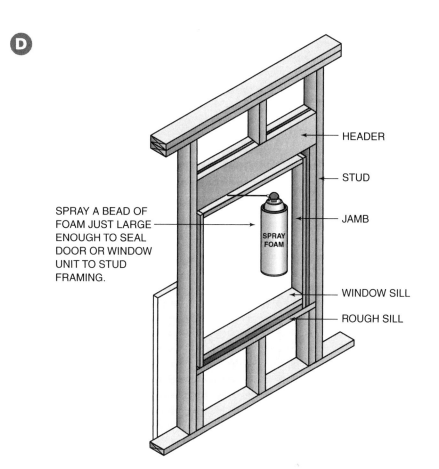

E Flexible insulation installed between floor joists over crawl spaces may be held in place by wire mesh or pieces of heavy-gauge wire wedged between the joists.

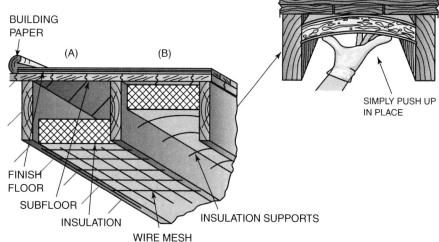

In (A), wire mesh is stapled to the edges of the joists.
In (B), pieces of heavy gauge wire, pointed at each end, are wedged between the joists to support the insulation.

Procedures: Installing Windows

A Place window in the opening after removing all shipping protection from the window unit. Do not remove any diagonal braces applied at the factory. Close and lock the sash. Windows can easily be moved through the openings from the inside and set in place.

B Center the unit in the opening on the rough sill with the exterior window casing against the wrapped wall sheathing. Level the window sill with a wood shim tip between the rough sill and the bottom end of the window's side jamb, if necessary. Secure the shim to the rough sill.

- Remove the window unit from the opening and caulk the backside of the casing or nailing flange. This will seal the unit to the building. Replace the unit and nail the lower end of both sides of the window. Next, plumb a side and nail the unit along the sides and top. Check that the sash operates properly. If not, make necessary adjustments.

C Flash the head casing by cutting to length the flashing with tin snips. Its length should be equal to the overall length of the window head casing. If the flashing must be applied in more than one piece, lap the joint about 3 inches. Slice the housewrap just above the head casing and slip the flashing behind the wrap and on top of the head casing. Secure with fasteners into the wall sheathing. Refasten the house wrap. Tape all seams in housewrap and over window nailing flanges.

A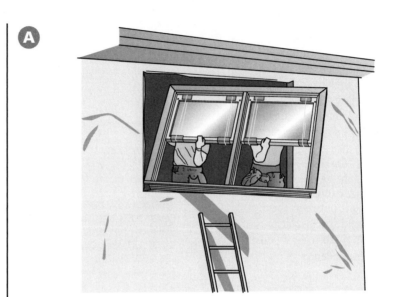

CAUTION Have sufficient help when setting large units. Handle them carefully to avoid damaging the unit or breaking the glass. Broken glass can cut through protective clothing and cause serious injury.

B **C**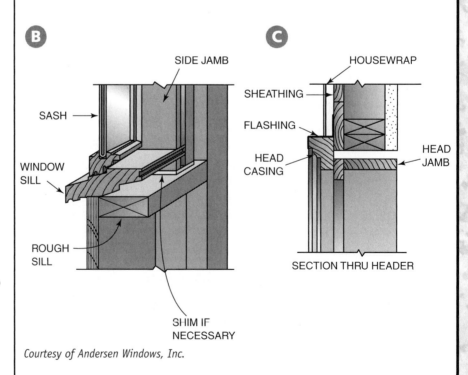

Courtesy of Andersen Windows, Inc.

Exterior Carpentry Maintenance

Procedures — Installing Gutters

A On both ends of the fascia, mark the location of the bottom side of the gutter. The top outside edge of the gutter should be in relation to a straight line projected from the top surface of the roof. The height of the gutter depends on the pitch of the roof.

- Stretch a chalk line between the two marks. Move the center of the chalk line up enough to give the gutter the proper pitch from the center to the ends. Snap a line on both sides of the center.

- Fasten the gutter brackets to the chalk line on the fascia with screws. All screws should be made of stainless steel or other corrosion resistant material. Aluminum brackets may be spaced up to 30 inches OC.

B Locate and install the outlet tubes in the gutter as required, keeping in mind that the downspout should be positioned plumb and square with the building. Add end caps and caulk all seams on the inside surfaces only.

- Hang the gutter sections in the brackets. Use slip-joint connectors to join larger sections. Use either inside or outside corners where gutters make a turn. Caulk all inside seams.

- Fasten downspouts to the wall with appropriate hangers and straps. Downspouts should be fastened at the top and bottom and every 6 feet in between. The connection between the downspout and the gutter is made with elbows and short straight lengths of downspout.

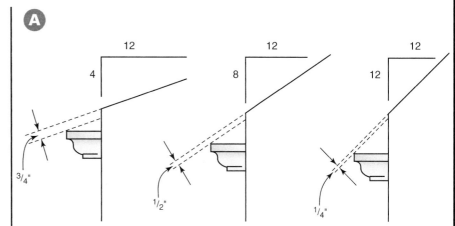

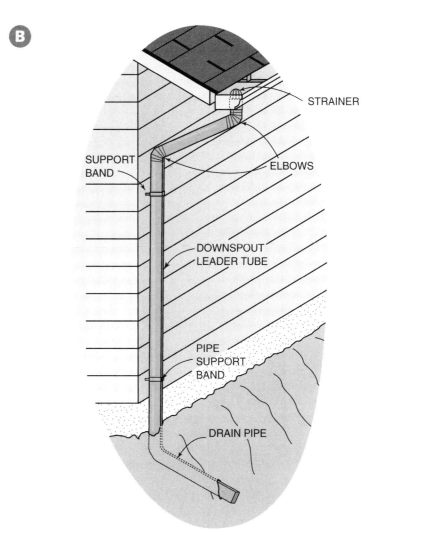

Procedures: Installing Gutters (Continued)

C Because water runs downhill, care should be taken when putting the downspout pieces together. The downspout components are assembled where the upper piece is inserted into the lower one. This makes the joint lap such that the water cannot escape until it leaves the bottom-most piece.

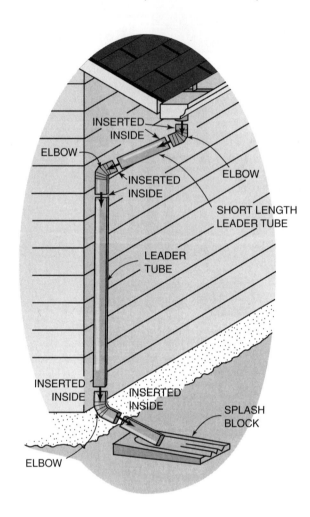

Procedures: Installing Asphalt Shingles

> **CAUTION:** Installation of roofing systems involve working on ladders and scaffolding as well as on top of the building. Workers should always be aware of the potential for falling. Keep the location of roof perimeter in mind at all times.

A Prepare the roof deck by clearing sawdust and debris that will cause a slipping hazard.

- Begin underlayment over the deck at a lower corner. Lap the following courses of felt over the lower course at least 2 inches. Make any end laps at least 4 inches. Lap the felt 6 inches from both sides over all hips.

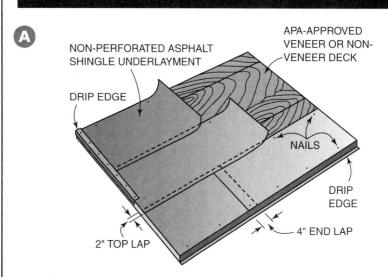

Courtesy of Asphalt Roofing Manufacturers Association.

B Nail or staple through each lap and through the center of each layer about 16 inches apart. Roofing nails driven through the center of metal discs or specially designed, large head felt fasteners hold the underlayment securely in strong winds until shingles are applied.

- Install metal drip edge along the perimeter on top of the underlayment. This will help prevent blow-offs.

- Prepare the starter course by cutting off the exposure taps lengthwise through the shingle. Save these tabs as they may be used as the last course at the ridge. Install the course so that no end joint will fall in line with an end joint or tab cutout of the regular first course of shingles.

Courtesy of APA—The Engineered Wood Association.

Procedures: Installing Asphalt Shingles (Continued)

FROM EXPERIENCE

Use a utility knife to cut shingles from the back side. Cut only half way through and then fold and break the shingle to complete the cut. When cutting from the granular top surface, use a hook blade.

C Determine the starting line, either the rake edge or vertical center snapped lines. To start from the middle of the roof, mark the center of the roof at the eaves and at the ridge. Snap a chalk line between the marks. Snap a series of chalk lines from this one, 4 or 6 inches apart, depending on the desired end tab, on each side of the centerline. When applying the shingles, start the course with the end of the shingle to the vertical chalk line. Start succeeding courses in the same manner. Break the joints as necessary, working both ways toward the rakes.

C
- SNAPPED LINES PERPENDICULAR TO FASCIA OR PARALLEL TO RAKE FASCIA
- METAL DRIP EDGE
- FIRST SHINGLE OF EACH COURSE STARTS AGAINST CHALK LINE
- STARTER STRIP

D Starting shingle layout at the rake edge involves placing the first course, with a whole tab at the rake edge. The second course is started with a shingle that is 6 inches shorter; the third course, with a strip that is a full tab shorter; the fourth, with one and one-half tabs removed, and so on. These starting pieces are precut for faster application.

D
- METAL DRIP EDGE APPLIED OVER FELT ALONG RAKE
- NAILING
- 2" HEAD LAP
- UNDERLAYMENT
- WOOD DECK
- EAVES FLASHING STRIP
- 1"
- 5 5/8"
- 4" END LAP
- SELF-SEALING STRIP
- METAL DRIP EDGE
- ① STARTER — BEGIN WITH A FULL STARTER SHINGLE MINUS 3" SO BUTT SEAMS DO NOT ALIGN WITH FIRST COURSE
- ② START FIRST COURSE WITH FULL STRIP
- ③ START SECOND COURSE WITH FULL STRIP MINUS 1/2 TAB
- ④ START THIRD COURSE WITH FULL STRIP MINUS FIRST TAB

E If the cutouts are to break on the thirds, cut the starting strip for the second course by removing 4 inches. Remove 8 inches from the strip for the third course, and so on.

- Fasten each shingle from the end nearest the shingle just laid. This prevents buckling. Drive fasteners straight so that the nail heads will not cut into the shingles. Both ends of the course should overhang the drip edge ¼ to ⅜ inch.

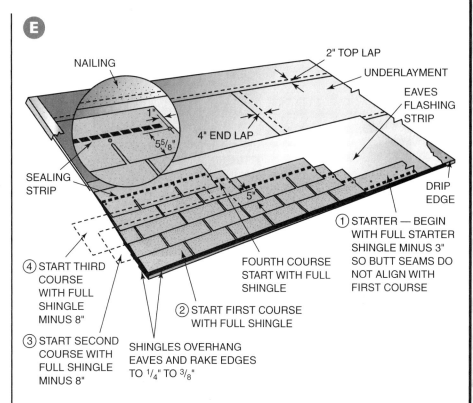

F Install vented ridge cap as per manufacturers instructions. Cut cap shingles and begin installation from one end. Center the cap shingle over the vented ridge cap. Secure each shingle with one fastener on each side.

- Apply the cap across the ridge until 3 or 4 feet from the end. Then space the cap to the end in the same manner as spacing the shingle course to the ridge. The last ridge shingle is cut to size. It is applied with one fastener on each side of the ridge. The two fasteners are covered with asphalt cement to prevent leakage.

Procedures: Installing Roll Roofing

Roll Roofing with Concealed Fasteners

A Apply 9-inch wide strips of the roofing along the eaves and rakes overhanging the drip edge about ³⁄₈ inch. Fasten with two rows of nails one inch from each edge spaced about 4 inches apart.

- Apply the first course of roofing with its edge and ends flush with the strips. Secure the upper edge with nails staggered about 4 inches apart. Do not fasten within 18 inches of the rake edge.

- Apply cement only to the edge strips covered by the first course. Press the edge and rake edges firmly to the strips. Complete the nails in the upper edge out to the rakes.

- Apply succeeding courses in like manner. Make all end laps 6 inches wide. Apply cement the full width of the lap.

- After all courses are in place, lift the lower edge of each course. Apply the cement in a continuous layer over the full width and length of the lap. Press the lower edges of the upper courses firmly into the cement. A small bead should appear along the entire edge of the sheet. Care must be taken to apply the correct amount of cement.

- To cover the hips and ridge, cut strips of 12" × 36" roofing. Bend the pieces lengthwise through their centers.

- Snap a chalk line on both sides of the hip or ridge down about 5½ inches from the center. Apply cement between the lines. Fit the first strip over the hip or ridge.

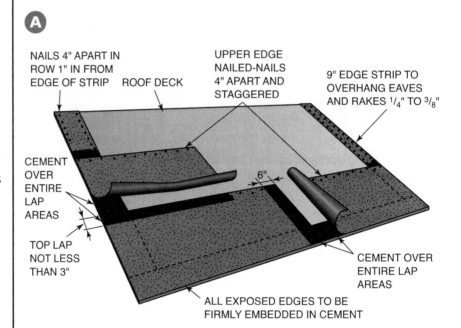

A

B Press it firmly into place. Start at the lower end of a hip and at either end of a ridge. Lap each strip 6 inches over the preceding one. Nail each strip only on the end that is to be covered by the overlapping piece.

- Spread cement on the end of each strip that is lapped before the next one is applied. Continue in this manner until the end is reached.

Double Coverage Roll Roofing

A Cut the 19-inch strip of *selvage*, non-mineral surface side, from enough double coverage roll roofing to cover the length of the roof. Save the surfaced portion for the last course at the ridge. Apply the selvage portion parallel to the eaves. It should overhang the drip edge by 3/8 inch. Secure it to the roof deck with three rows of nails.

B Apply the first course using a full width strip of roofing. Secure it with two rows of nails in the selvage portion.

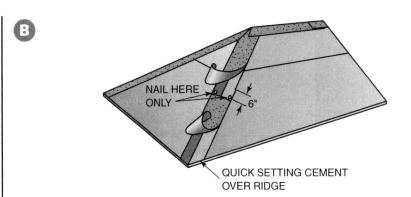

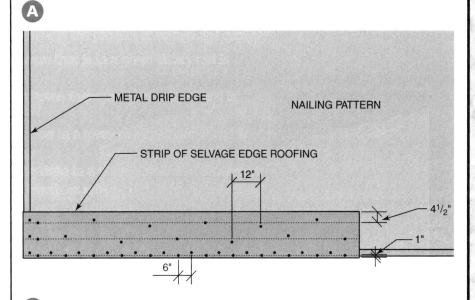

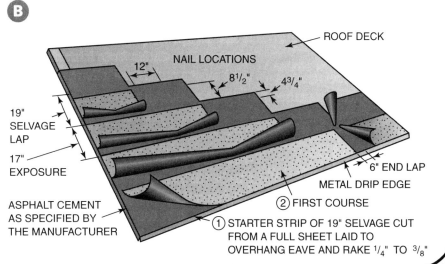

Procedures: Installing Roll Roofing (Continued)

C Apply succeeding courses in the same manner. Lap the full width of the 19-inch selvage each time. Make all end laps 6 inches wide. End laps are made in the manner shown in the accompanying figure. Stagger end laps in succeeding courses.

- Lift and roll back the surface portion of each course. Starting at the bottom, apply cement to the entire selvage portion of each course. Apply it to within ¼ inch of the surfaced portion. Press the overlying sheet firmly into the cement. Apply pressure over the entire area using a light roller to ensure adhesion between the sheets at all points.

D Apply the remaining surfaced portion left from the first course as the last course. Hips and ridges are covered in the same manner shown in the accompanying figure.

- It is important to follow specific application instructions because of differences in the manufacture of roll roofing. Specific requirements for quantities and types of adhesive must be followed.

C
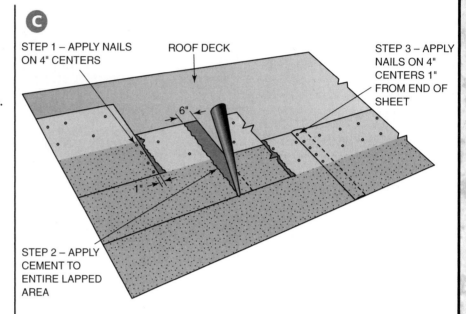

D
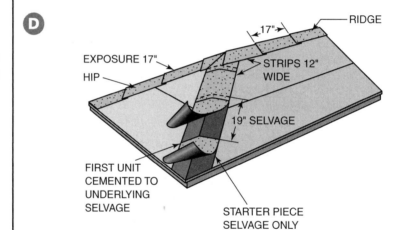

Procedures

Woven Valley Method

A Install underlayment and starter strip to both roofs.

- Apply first course of one roof, say the left one, into and past the center of the valley. Press the shingle tightly into the valley and nail, keeping the nails at least 6 inches away from the valley centerline. Cut shingles to adjust the butt ends so there is no butt seam within 12 inches of the valley centerline.

- Apply the first course of the other (right) roof in a similar manner, into and past the valley.

- Succeeding courses are applied by repeating this alternating pattern, first from one roof and then on the other.

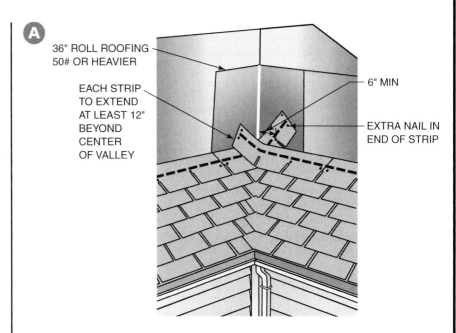

Procedures

Closed Cut Valley Method

A Begin by shingling first roof completely, letting the end shingle of every course overlap the valley by at least 12 inches. Form the end shingle of each course snugly into the valley. Cut shingles to adjust the butt ends so there is no butt seam within 12 inches of the valley centerline.

- Snap a chalk line along the center of the valley on top of the shingles of the first roof.

- Apply the shingles of second roof, cutting the end shingle of each course to the chalk line. Place the cut end of each course that lies in the valley in a 3-inch wide bed of asphalt cement.

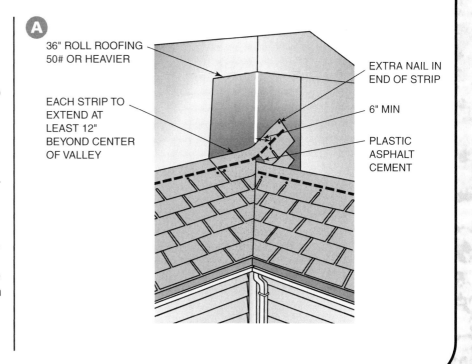

Procedures: Step Flashing Method

A Snap a chalk line in the center of the valley on the valley underlayment.

- Apply the shingle starter course on both roofs. Trim the ends of each course that meet the chalk line.

- Fit and form the first piece of flashing to the valley on top of the starter strips. Trim the bottom edge flush with the drip edge. Fasten with two nails in the upper corners of the flashing only. Use nails of like material to the flashing to prevent electrolysis.

- Apply the first regular course of shingles to both roofs on each side of the valley, trimming the ends to the chalk line. Bed the ends in plastic asphalt cement. Do not drive nails through the metal flashing. Apply flashing to each succeeding course in this manner.

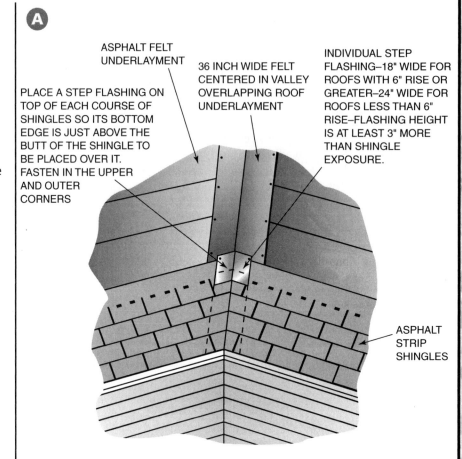

A

PLACE A STEP FLASHING ON TOP OF EACH COURSE OF SHINGLES SO ITS BOTTOM EDGE IS JUST ABOVE THE BUTT OF THE SHINGLE TO BE PLACED OVER IT. FASTEN IN THE UPPER AND OUTER CORNERS

ASPHALT FELT UNDERLAYMENT

36 INCH WIDE FELT CENTERED IN VALLEY OVERLAPPING ROOF UNDERLAYMENT

INDIVIDUAL STEP FLASHING—18" WIDE FOR ROOFS WITH 6" RISE OR GREATER—24" WIDE FOR ROOFS LESS THAN 6" RISE—FLASHING HEIGHT IS AT LEAST 3" MORE THAN SHINGLE EXPOSURE.

ASPHALT STRIP SHINGLES

Procedures — Installing Horizontal Siding

 First determine the siding exposure so that it is about equal both above and below the window sill. Divide the overall height of each wall section by the maximum allowable exposure. Round up this number to get the number of courses in that section. Then divide the height again by the number of courses to find the exposure. These slight adjustments in exposure will not be noticeable to the eye.

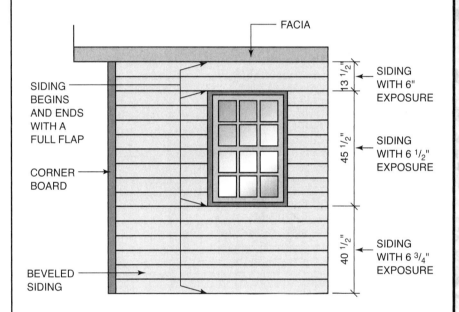

EXAMPLE: Consider the overall dimensions in the accompanying figure. Divide the heights by the maximum allowable exposure, 7 inches in this example. Then round up to the nearest number of courses that will cover that section. Divide the section height by the number of courses to find the exposure.

$40\frac{1}{2} \div 7 = 5.8 \Rightarrow 6$ courses $40\frac{1}{2} \div 6 = 6.75$ or $6\frac{3}{4}$" exposure
$45\frac{1}{5} \div 7 = 6.5 \Rightarrow 7$ courses $45\frac{1}{2} \div 7 = 6.5$ or $6\frac{1}{2}$" exposure
$12\frac{1}{2} \div 7 = 1.8 \Rightarrow 2$ courses $12\frac{1}{2} \div 2 = 6.25$ or $6\frac{1}{4}$" exposure

Procedures: Installing Horizontal Siding (Continued)

B Install a starter strip of the same thickness and width of the siding at the headlap fastened along the bottom edge of the sheathing. For the first course, a line is snapped on the wall at a height of the top edge of the first course of siding.

- From this first chalk line, layout the desired exposures on each corner board and each side of all openings. Snap lines at these layout marks. These lines represent the top edges of all siding pieces.

- Install the siding as per manufacturer's recommendations, staggering the butt joints in adjacent courses as far apart as possible. A small piece of felt paper is used behind the butt seams to ensure the weathertightness of the siding.

C When applying a course of siding, start from one end and work toward the other end. With this procedure, only the last piece will need to be fitted. Tight-fitting butt joints must be made between pieces. Measure carefully and cut it slightly longer. Place one end in position. Bow the piece outward, position the other end, and snap into place. Take care not to move the corner board with this technique. Do not use this technique on cementitious siding.

D Siding is fastened to each bearing stud or about every 16 inches. On bevel siding, fasten through the butt edge just above the top edge of the course below. Do not fasten through the lap. This allows the siding to swell and shrink with seasonal changes without splitting of the siding. Cementitious siding may be blind nailed by fastening along the top edge only. Blind nailing is not recommended in high-wind areas.

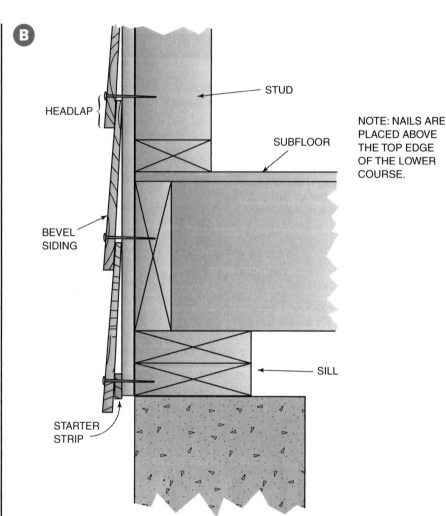

SECTION THROUGH SILL

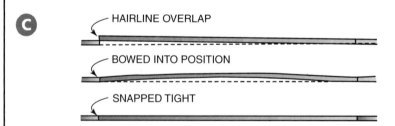

Courtesy of California Redwood Association.

 FROM EXPERIENCE

Setting up a comfortable work station for cutting will allow the carpenter to work more efficiently and safely. This will also reduce waste and improve workmanship.

CHAPTER 7 *Carpentry* **189**

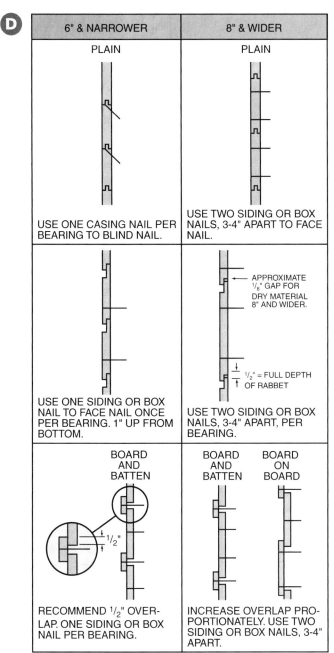
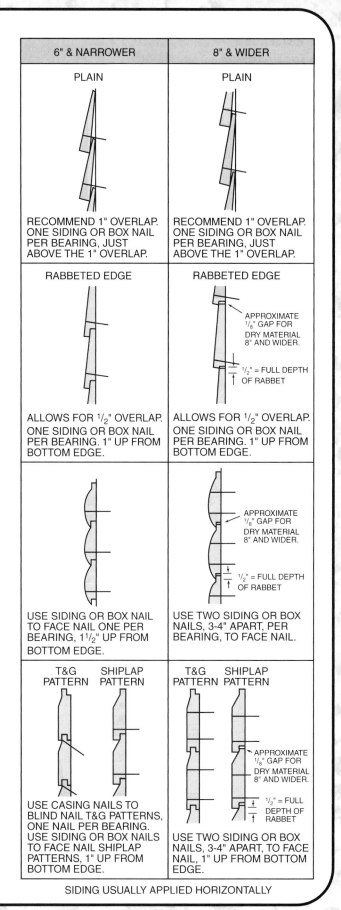

Courtesy of Western Wood Products Association.

Procedures: Installing Vertical Tongue-and-Groove Siding

A Slightly back-bevel the ripped edge. Place it vertically on the wall with the beveled edge flush with the corner similar to making a corner board. Face nail the edge nearest the corner.

- Fasten a temporary piece on the other end of the wall projecting below the sheathing by the same amount. Stretch a line to keep the bottom ends of other pieces in a straight line.

B Apply succeeding pieces by toe-nailing into the tongue edge of each piece. Make sure the edges between boards come up tight. Drive the nail home until it forces the board up tight. Make sure to keep the bottom ends in a straight line. If butt joints are necessary, use a mitered or rabbeted joint for weathertightness.

C To cut the piece to fit around an opening, first fit and tack a siding strip in place where the last full strip will be located. Level from the top and bottom of the window casing to this piece of siding. Mark the piece.

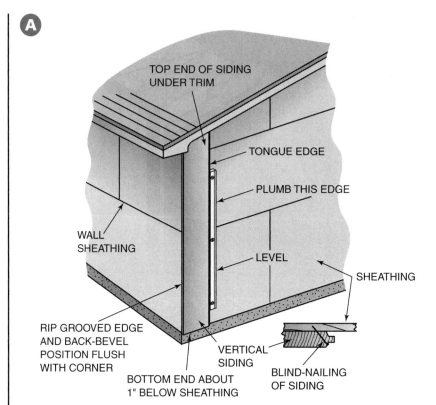

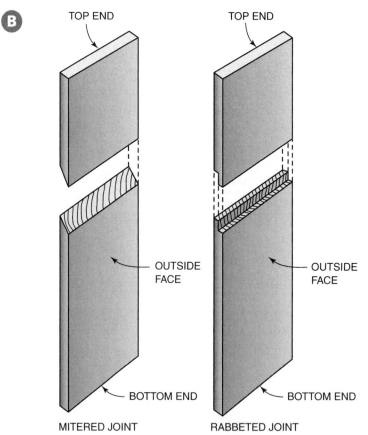

MITERED JOINT   RABBETED JOINT

- Next, use a scrap block of the siding material, about 6 inches long, with the tongue removed. Be careful to remove all of the tongue, but no more. Hold the block so its grooved edge is against the side casing and the other edge is on top of the tacked piece of siding. Mark the vertical line on the siding by holding a pencil against the outer edge of the block while moving the block along the length of the side casing. Remove and cut the piece, following the layout lines carefully. Set this piece aside for the time being. Cut and fit another full strip of siding in the same place as the previous piece. Fasten both pieces in position.

- Continue the siding by applying the short lengths across the top and bottom of the opening as needed.

- **D** Fit the next full-length siding piece to complete the siding around the opening. First tack a short length of siding scrap above and below the window and against the last pieces of siding installed. Tack the length of siding to be fitted against these blocks in the grooves. Level and mark from the top and bottom of the window to the full piece. Lay out the vertical cut by using the same block with the tongue removed, as used previously. Hold the grooved edge against the side casing. With a pencil against the other edge, ride the block along the side casing while marking the piece to be fitted.

- Remove the piece and the scrap blocks from the wall. Carefully cut the piece to the layout lines. Then fasten in position. Continue applying the rest of the siding.

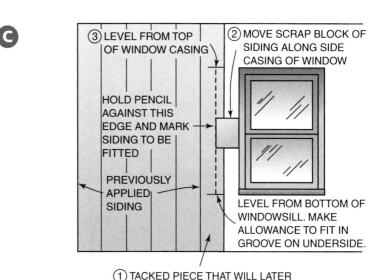

Procedures

Installing Panel Siding

A Install the first piece with the vertical edge plumb. Rip the sheet to size, putting the cut edge at the corner. The factory edge should fall on the center of a stud. Panels must also be installed with their bottom ends in a straight line. It is important that horizontal butt joints be offset and lapped, rabbeted, or flashed. Vertical joints are either shiplapped or covered with **battens.**

- Apply the remaining sheets in the first course in like manner. Cut around openings in a similar manner as with vertical tongue-and-grooved siding. Carefully fit and caulk around doors and windows. Trim the end of the last sheet flush with the corner.

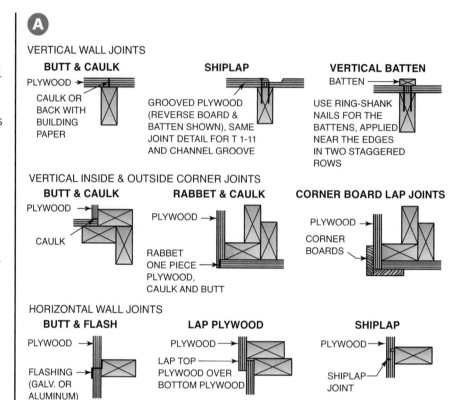

Courtesy of APA—The Engineered Wood Association.

Procedures

Installing Wood Shingles and Shakes

A Fasten a shingle on both ends of the wall with its butt about 1 inch below the top of the foundation. Stretch a line between them from the bottom ends. Fasten an intermediate shingle to the line to take any sag out of the line. Even a tightly stretched line will sag in the center over a long distance.

- Fill in the remaining shingles to complete the undercourse. Take care to install the butts as close to the line as possible without touching it. Remove the line.

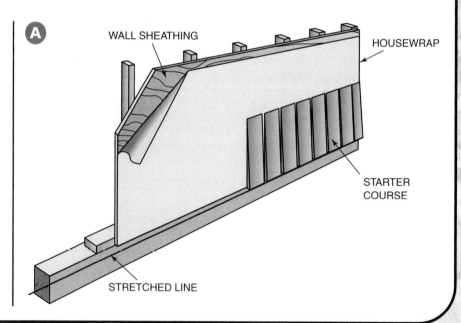

Procedures

Installing Wood Shingles and Shakes (Continued)

B Apply another course on top of the first course. Offset the joints in the outer layer at least 1½ inches from those in the bottom layer. Shingles should be spaced ⅛ to ¼ inch apart to allow for swelling and to prevent buckling. Shingles can be applied close together if factory-primed or if treated soon after application.

B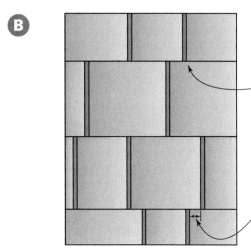

SHINGLES SPACED ⅛" TO ¼" APART. THESE JOINTS ALLOW FOR EXPANSION AND PREVENT POSSIBLE "BUCKING." FACTORY PRODUCTS MAY BE CLOSER.

LEAVE A SIDE LAP OF AT LEAST 1½" BETWEEN JOINTS IN SUCCESSIVE COURSES.

Courtesy of Cedar Shake and Shingle Bureau.

C To apply the second course, snap a chalk line across the wall at the shingle butt line. Using only as many finish nails as necessary, tack 1 × 3 straightedges to the wall with their top edges to the line. Lay individual shingles with their butts on the straightedge.

C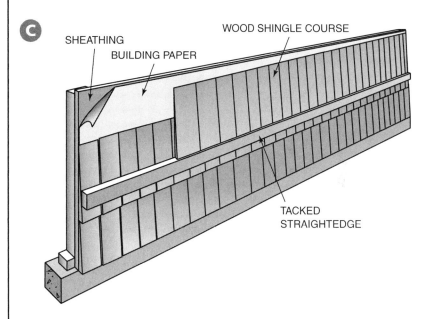

SHEATHING
BUILDING PAPER
WOOD SHINGLE COURSE
TACKED STRAIGHTEDGE

Procedures: Applying Horizontal Vinyl Siding

A Snap a level line to the height of the starter strip all around the bottom of the building. Fasten the strips to the wall with their edges to the chalk line. Leave a ¼-inch space between them and other accessories to allow for expansion. Make sure the starter strip is applied as straight as possible. It controls the straightness of entire installation.

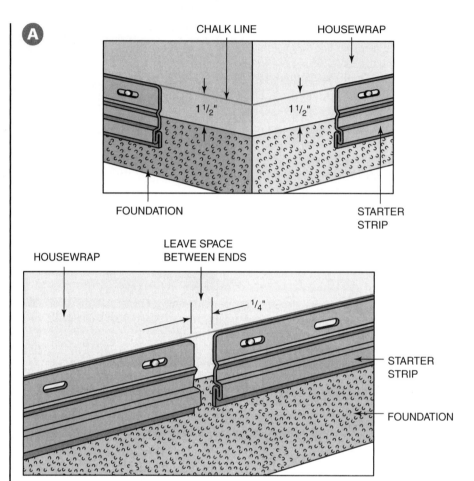

B Cut the corner posts so they extend ¼ inch below the starting strip. Attach the posts by fastening at the top of the upper slot on each side. The posts will hang on these fasteners. The remaining fasteners should be centered in the nailing slots. Make sure the posts are straight, plumb, and true from top to bottom.

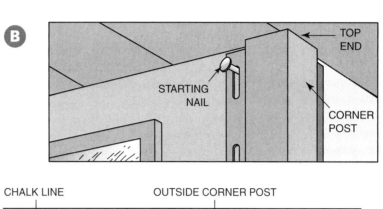

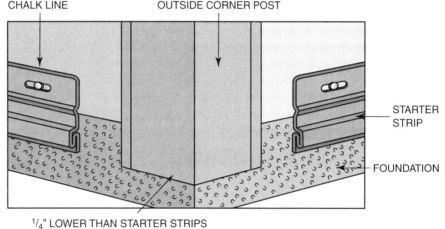

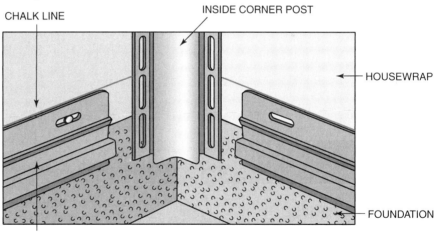

Procedures

Applying Horizontal Vinyl Siding (Continued)

C Cut each J-channel piece to extend, on both ends, beyond the casings and sills a distance equal to the width of the channel face. Install the side pieces first by cutting a ¾-inch notch, at each end, out of the side of the J-channel that touches the casing. Fasten in place.

- On both ends of the top and bottom channels, make ¾-inch cuts at the bends leaving the tab attached. Bend down the tabs and miter the faces. Install them so the mitered faces are in front of the faces of the side channels.

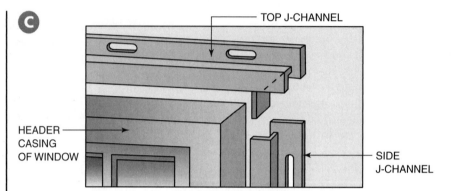

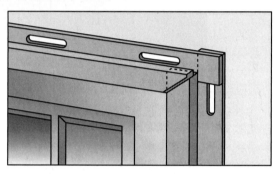

COMPLETE INSTALLATION WITH TOP J-CHANNEL ON OUTSIDE OF SIDE J-CHANNEL

D Snap the bottom of the first panel into the starter strip. Fasten it to the wall. Start from a back corner, leaving a ¼-inch space in the corner post channel. Work toward the front with other panels. Overlap each panel about 1 inch. The exposed ends should face the direction from which they are least viewed.

- Install successive courses by interlocking them with the course below and staggering the joints between courses.

E To fit around a window, mark the width of the cutout, allowing ¼-inch clearance on each side. Mark the height of the cutout, allowing ¼-inch clearance below the sill. Using a special *snaplock punch*, punch the panel along the cut edge at 6-inch intervals to produce raised lugs facing outward. Install the panel under the window and up in the undersill trim. The raised lugs cause the panel to fit snugly in the trim.

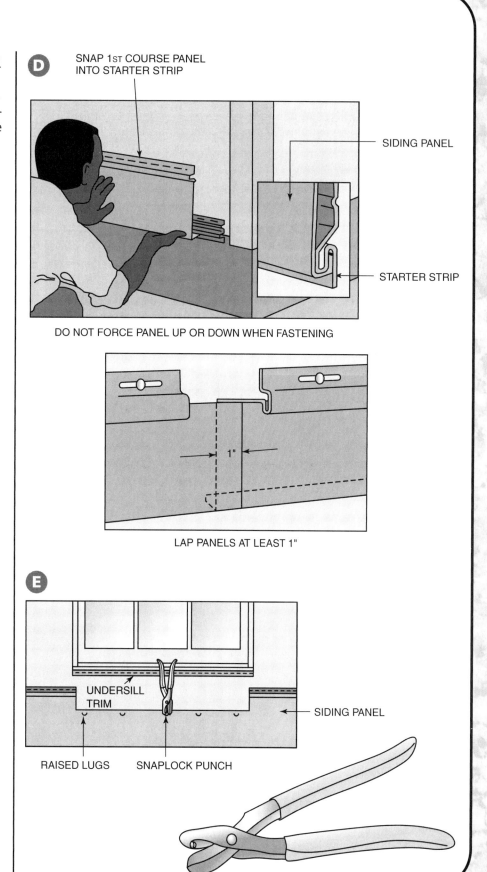

Procedures

Applying Horizontal Vinyl Siding (Continued)

F Panels are cut and fit over windows in the same manner as under them. However, the lower portion is cut instead of the top. Install the panel by placing it into the J-channel that runs across the top of the window.

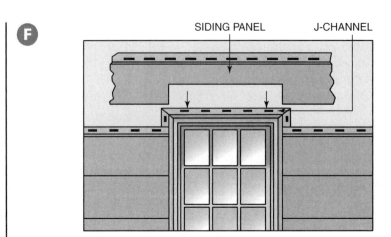

CUT EDGE OF PANEL FITS INTO J-CHANNEL OVER TOP OF WINDOW

G Install the last course of siding panels under the soffit in a manner similar to fitting under a window. An *undersill trim* is applied on the wall and up against the soffit. Panels in the last course are cut to width. Lugs are punched along the cut edges. The panels are then snapped firmly into place into the undersill trim.

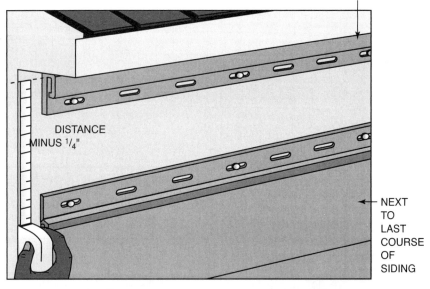

① MEASURE FOR LAST COURSE OF SIDING

② RAISED LUGS

③ INSTALL CUT EDGE INTO UNDERSILL TRIM

Procedures: Applying Vertical Vinyl Siding

 Measure and lay out the width of the wall section for the siding pieces. Determine the width of the first and last piece.

- Cut the edge of the first panel nearest the corner. Install an undersill trim in the corner board or J-channel with a strip of furring or backing. This will keep the edge in line with the wall surface. Punch lugs along the cut edge of the panel at 6-inch intervals. Snap the panel into the undersill trim. Place the top nail at the top of the nail slot. Fasten the remaining nails in the center of the nail slots.

- Install the remaining full strips making sure there is ¼-inch gap at the top and bottom. Fit around openings in the same manner as with fitting vertical siding. Install the last piece into undersill trim in the same manner as for the first piece.

EXAMPLE: What is the starting and finishing widths for a wall section that measures 18'–9" for siding that is 12" wide?

Convert this measurement to a decimal by first dividing the inches portion by 12 and then adding it to the feet to get 18.75'.

Divide this by the siding exposure, in feet: 18.75 ÷ 1 foot = 18.75 pieces.

Subtract the decimal portion along with one full piece giving 1.75 pieces. Next 1.75 ÷ 2 = 0.875, multiplied by 12 gives 10½".

This is the size of the starting and finishing piece. Thus there are 17 full width pieces and two 10½" wide pieces.

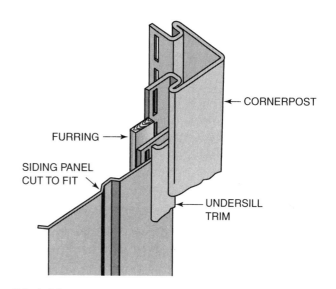

Courtesy of Vinyl Siding Institute.

Review Questions

Select the most appropriate answer.

1. True or false? The most common method of classifying wood is by its source.

2. List three common hardwoods.

3. List three common softwoods.

4. Define the term engineer panels.

5. List the steps for estimating the amount of drywall needed for a particular application.

Name: _____

Date: _____

Carpentry Job Sheet

1

Carpentry

Identify Characteristics of Wood

Upon completion of this job sheet, you should be able to understand how wood is classified.

① Go to your local lumber yard or building supply center where there is a variety of different types of wood. Make a list of the different types of wood available and classify it as either hardwood or softwood.

Name of the Wood	Classification

② Are the woods available native to your area? (Y/N) If not, where does the wood come from?

③ At your local lumber yard or building supply center, list the types of engineered panels they make available.

④

Type of Panel	How They Are Made

Instructor's Response:

_____
_____
_____
_____

Chapter 8 Surface Painting

OBJECTIVES

By the end of this chapter, you will be able to:

Knowledge-Based

- Identify and select proper surface finishes.
- Identify and select proper finishing tools for different types of finishes.

Skill-Based

- Prepare surface and site properly for finishing, including sanding, caulking, and covering exposed surfaces.
- Apply paint by using roller and brush according to manufacturer and job specifications.
- Apply paint by using a paint sprayer according to manufacturer and job specifications.
- Clean and store paint materials including brushes, rollers, thinners, and spray guns according to manufacturer's specifications and OSHA regulations.

Introduction

The proper application of paint can change the look of a room at very little cost. Using the right paint, selecting the proper products and equipment, and using professional-style techniques will give you the results worthy of any room. You will also learn how to prepare a wall for excellent results and to use and clean your tools correctly.

Surface Preparation

Before starting the actual painting, it is important to prepare the surfaces properly to get a professional-looking paint job.

Identify and Select Proper Surface Finishes

To start the paint process, select the proper paint for the type of surface, location of the surface, and use of the area.

Types of Paint

There are many different types of paint, each designed for a specific purpose, surface, or place in a building.

- Latex paints—water-based, durable, and lower in odor than oil-based products. In addition, latex paints can be cleaned up easily with soap and water. For these reasons, latex is preferred for most interior surfaces. Oil paints are durable and resist scraping and wear and tear. Most contain alkyd, a Soya-based resin that dries harder than latex. Because these paints are solvent-based, brushes and spills must be cleaned up using paint thinner.
- Gloss paints—oil-based and include resin to give them a hard-wearing quality. The higher the gloss level, the higher the shine and easier it is to maintain.
- Flat paints—ideal for low-traffic areas such as formal dining rooms and master bedrooms. They provide a beautiful matte coating that hides minor surface imperfections.
- Eggshell paints—provide a smooth finish with a subtle sheen that is slightly glossier than flat. It is washable, durable, and ideal for bedrooms, hallways, home offices, and family rooms.
- Satin paints—a step above eggshell in ability to be cleaned, providing a nice balance between being washable and having a subtle gloss. They perform and look great in just about any room.
- Semi-gloss paints—ensure maximum durability. They are commonly used in children's rooms and high-moisture areas such as bathrooms, as well as for trim.
- High-gloss paints—highly reflective and work well for highlighting details, such as trim and decorative molding. They are also the best choice for doors, cabinets, or any area that sees a high volume of abuse.
- Ceiling flats—designed specially for ceilings. These are usually extra spatter-resistant.

Identify and Select Proper Finishing Tools for Type of Finish

Selecting the proper tools will save you a lot of time and additional effort. Always choose top-quality tools.

Types of Finishing Tools

Pressure washers—will blast away many years worth of dirt and grime in a short time and with minimal amount of effort.

Paint scrapers—tools used to efficiently remove the loose and peeling paint from any surface. There are many types of paint scrapers. Examples include the classic putty knife (Figure 8-1), the 11-in-1 multipurpose tool, and the double-edge wood scraper.

Power sanders—help with the removal of old peeling paint.

Figure 8-1: **Putty knife** (©2006 JupiterImages Corporation)

Paint brushes—choosing the right brush for your task will make the job easier and provide the best results. Determining which paint brush is best for your project is based on type of paint or finish, surface to be painted, and which one you are most comfortable in using.

Types of Brushes

There are two different categories of brushes: water-based paint brushes and oil-based paint brushes. It is important to choose the appropriate type of brush for the type of finish and surface.

Water-Based Paint Brushes

There are three main types of water-based paint brushes (see Figure 8-2).

Figure 8-2: **Polyester, nylon, and poly/nylon paint brushes**

- **Polyester**—these bristle brushes hold and release more paint. This provides a smoother finish. Clean-up is faster and more thorough than with other synthetic brushes.
- **Nylon**—These bristle brushes wear longer and are stiffer than any other filament. A nylon brush is well suited for rough surfaces.
- **Poly/nylon**—These blends provide longer wear, maximum resiliency, and easy clean-up.

Figure 8-3: **White China bristle and Black China bristle brushes**

Oil-Based Paint Brushes

There are two basic types of oil-based paint brushes (see Figure 8-3).

- **White China bristle**—the best choice for varnishes, polyurethane, and stains. White China bristles are finer than Black China bristles and provide a finer finish.
- **Black China bristle**—best used with oil-based paint, primer, and enamels.

Brush Sizes and Shapes

Paint brushes come in a variety of shapes, such as angular, flat, and oval. Sizes range from 1 inch to 4 inches in width (see Figure 8-4).

- **Angular brushes**—great for angular or narrow surfaces. They are also good brushes to use when the painting surface is hard to reach. An angular brush is an excellent choice for an all-purpose brush.
- **Flat and oval brushes**—can be used on all surfaces, but are best suited for flat surfaces such as wide trim, doors, cutting-in walls, or ceilings.

Figure 8-4: **Various brush sizes** (©2006 JupiterImages Corporation)

Roller Covers

The most important factor in selecting a paint roller cover is the surface that is going to be painted. The rule for using almost all roller covers is as follows: The smoother the surface to be painted, the shorter the nap; the rougher the surface, the longer the nap. A high-quality roller cover should have a phenolic core that will not soften in water and that will withstand every type of paint solvent. Rollers also are available in natural materials (mohair or lamb's wool) and synthetic materials (nylon, polyester, or a combination of the two.) Natural materials are best with oil-based paints. Synthetic materials are best with water-based paints. For latex paint, only use synthetic rollers. Natural materials are too absorbent (see Figure 8-5).

- **Smooth surface**—select a short nap (1/8-inch to 1/4-inch) cover for smooth surfaces. A longer nap can leave a pronounced "orange peel" effect. Use this short nap cover on smooth plaster, sheet rock, wallboard, smooth wood, Masonite, and Celotex.
- **Slightly rough surface**—select a medium nap (3/8-inch to 1/2-inch) cover for slightly rough surfaces. Longer fibers push the paint into rough surfaces without causing an "orange peel" effect. Use this medium nap cover on sand finish plaster, texture plaster, acoustical tile, poured concrete, rough wood, and shakes.
- **Rough surface**—select a long nap (3/4-inch to 1 1/4-inch) cover for rough surfaces. The longer fibers push paint into the deep valleys of rough surfaces. Use this cover on concrete block, stucco, brick, Spanish plaster, cinder block, corrugated metal, and asphalt or wood shingles.

Roller Frames

Roller cage frames come in a variety of styles (see Figure 8-6). U-shaped frames are generally sturdier. When choosing frames, be sure to select those that are sealed on the ends to help keep the paint on the roller, where it belongs.

Figure 8-5: Paint roller and pan (©2006 JupiterImages Corporation)

Figure 8-6: Various roller cage frame types

Properly Prepare the Surface for Painting

Surface preparation is a critical first step if you want a good-quality paint job. The best paint money can buy will not help if the painting surface is not ready to accept it.

Steps for Preparing the Surface for Painting

1. **Remove as much furniture as possible from the room.** Group the heavier items and cover them with drop cloths. Protect the baseboards with 12-inch baseboard masking and a drop cloth (Figure 8-7).
2. **Repair drywall.** Drywall problems are relatively easy to spot. Nails sometimes pop out from the drywall. Corners get dented, scraped, or otherwise damaged. Tape can split. Dents, gouges, and holes appear. All drywall problems are relatively easy to fix. See "Fixing Drywall Problems" on page 208.
3. **Remove switch plates and receptacle plates.** Never try to tape or paint over switch plates or receptacle plates. Always remove them. Be sure to tape the screws to the plate so that they are not lost during the painting (Figure 8-8).
4. **Caulk and cover exposed surfaces.** Caulk all joints, cracks, and seams in the surface before painting. This can be done with a caulking gun and caulking cartridges (Figure 8-9).
5. **Check around the windowpanes for loose putty.** If putty is loose or missing, replace with new putty before starting the painting job.

Figure 8-7: Furniture covered with drop cloth

Figure 8-8: Switch plate has been removed (©2006 JupiterImages Corporation)

Figure 8-9: Caulking gun and cartridge

Safety Tip

Be extremely careful when moving furniture to protect yourself and the furniture.

- If you know you must move large furniture, get a helper.
- Use a back support to protect your back.
- Walk large, heavy pieces by moving legs to the left and right in small increments.
- Turn a piece of carpet upside down to make it easier to slide.
- Use a dolly to move heavy pieces.

Fixing Drywall Problems

Figure 8-10: Wall with nail holes

Drywall can pose several problems that must be repaired before starting the painting process. Drywall can have nail pops, which are visible dimples in the drywall caused by underlying nails. Drywall can also have small holes caused by small nails used for hanging pictures, thumbtacks, or small dents caused by furniture. In some cases, you may see large holes in the drywall that must be patched properly to ensure a smooth paint result (Figure 8-10).

Steps for Fixing Nail Pops

1. **If you can remove the nail without damage, do so.** Refasten the drywall with drywall screws. Drive the screws until they are recessed but do not break through the paper covering on the drywall (Figure 8-11).
2. **If the nail cannot be removed, drive it back in place.** Fill the dent and screw holes with joint compound (Figure 8-12).

Figure 8-11: Using drywall screws

Figure 8-12: Repairing flaws in the wall (©2006 JupiterImages Corporation)

Steps for Repairing a Small Hole in Drywall

1. **Remove loose drywall plaster and cut away torn paper with a utility knife** (Figure 8-13).
2. **Roughen the edges of the hole with coarse sandpaper. Wipe dust away from the hole** (Figure 8-14).

Figure 8-13: Utility knife (©2006 JupiterImages Corporation)

Figure 8-14: Rough up hole with sandpaper

Figure 8-15: Putting repair in place

Figure 8-16: Cover screen with compound

3. **Patch the hole.** Cut a piece of wire screening slightly larger than the hole or use a drywall patch screen and cover the hole with it. You may need a coat of fresh compound or a string to keep the screen in place (Figure 8-15).
4. **Cover the screen with compound.** Let it dry before continuing (Figure 8-16).
5. **Sand, prime, and paint** (Figure 8-17).

Steps for Patching a Large Drywall Hole

1. **Mark out a rectangle around the hole with a straightedge or carpenter's rule** (Figure 8-18).
2. Cut through the paper surface on the marked lines by using a utility knife or keyhole saw (Figure 8-19).
3. Cut a drywall patch.
   Cut a drywall patch two inches each direction larger than the hole. Remove the 2-inch perimeter, but leave the facing paper (Figure 8-20).
4. **Spread joint compound around the outside edges of the hole and along its inside edges** (Figure 8-21).

Figure 8-17: Sand and paint repaired hole

Figure 8-18: Draw rectangle around hole

Figure 8-19: Cut rectangular hole in the wall

Figure 8-20: Cut drywall patch

Figure 8-21: Joint compound on patch

Figure 8-22: Position patch on hole being repaired

Figure 8-23: Paint patch

5. Place the patch in position and hold. Hold it in place for several minutes while it begins to adhere. Spread more joint compound as needed with a drywall knife. Allow the compound to dry completely (Figure 8-22).
6. **Sand, prime, and paint** (Figure 8-23).

Apply Paint by Using a Roller and a Brush

The proper sequence for painting a room is from top to bottom. Paint the ceiling first, then the walls, the windows, the doors, and the baseboards, in that order.

Painting the Ceiling

If the walls are going to be painted, there is no need to worry about protecting them. However, if you are painting the ceiling only, you should protect the walls with plastic drop cloths.

Steps for Painting the Ceiling

1. **Paint a 2-inch-wide strip with a brush around the edges of the ceiling.**
2. **Switch to a roller with a 4- to 5-foot extension pole.** Starting at a corner, paint a section about 3 feet square. Use a zigzag pattern. Paint a "W" pattern on the ceiling. This will disperse the paint on the roller evenly.
3. **Fill in this 3-foot section.** Paint the section without reloading the roller until you have completely covered the section.
4. **Continue to cover the ceiling.** Work across the ceiling's shortest dimension in 3-foot square sections. Overlap your strokes while the paint is wet to minimize lap marks (Figure 8-24).

Figure 8-24: **Paint ceiling in small sections**

Painting the Walls

Once the ceiling has been completed, move to the walls. Be sure that the walls are fully prepared to be painted.

Steps for Painting Walls

1. **Prime the walls.** Proper priming is key to a long-lasting job.
2. **Paint a 2-inch strip at the ceiling with a brush.**
3. **Paint 2-inch strips in corners, around windows, doors, cabinets, and baseboards.**
4. **Switch to a roller.** Paint in a vertical direction using a zigzag pattern. Push the roller upward on the first stroke and then form an "M" pattern to evenly distribute the paint on the roller. Work in 3-foot sections. Fill in the "M" pattern without reloading the roller until you have completely covered the area. Continue in 3-foot sections until the wall is finished. Touch up spots you missed when the paint is wet to help minimize sheen differences (Figure 8-25).

Figure 8-25: **Painting a wall with a roller**

Steps for Applying Paint with a Paint Sprayer

1. **Pour the paint through a strainer into the bin or bucket.** Be sure to avoid lumps or odd bits of non-paint material (Figure 8-26).
2. **Thin the paint.** Do not thin paint more than recommended by the manufacturer or it will not cover well.
3. **Cover yourself very well.** Be sure that you are covered from head to toe. Wear a long-sleeved shirt, gloves, and dust mask or respirator (Figure 8-27).
4. **Paint from the top down in smooth, steady strokes.** Start at a corner, work from the top down, and keep your strokes steady and smooth. A lot of paint is going on the surface in a short period of time. It is better to paint several light coats than one heavy one (Figure 8-28).

Figure 8-26: Paint strainer

Figure 8-27: Protected painter

Figure 8-28: Paint sprayer

Cleaning and Storing of Equipment and Supplies

It is important to clean all paint equipment thoroughly, store equipment and supplies properly, and discard empty containers appropriately.

Steps for Cleaning Paint Rollers

1. **Disassemble paint roller.**
2. **Submerge cover in solvent.**
3. **Wash and rinse cover.** Wash the roller cover in mild detergent. Rinse it in clear water (see Figure 8-29).
4. **Remove paint from frame and hardware with solvent.**
5. **Hang roller to dry.**

CHAPTER 8 *Surface Painting* 213

Figure 8-29: Rinse paint roller under tap

Figure 8-30: Person removing excess paint from brush with a scraper

Steps for Cleaning Paint Brushes

1. **Remove excess paint**. Either work it out on a piece of newspaper or run the edge of a scraper along the bristles (Figure 8-30).
2. **Use water to clean water-based paints.** Water-based paints can simply be rinsed out under a running tap. Work them in a bucket of water first to remove much of the paint build-up. This makes rinsing them under the tap faster and easier. Make sure the water running through the brush is clear before finishing (Figure 8-31).
3. **Use white spirit to clean oil-based paint.** Brushes used in oil-based paint should be worked in a container of white spirit first. Then, shake out the excess paint and work the brush in brush cleaner. Brush cleaner is water-soluble so the brush can then be rinsed under a running tap. Wash the brushes in soapy water and rinse under running water.

Figure 8-31: Clean brush with water

4. **Shake out the excess water by spinning the brush handle between your palms.**
5. **Wrap the brush in paper to retain its shape.** Tip: Paint that has dried into the brush can be removed by soaking it in brush restorer. Really dried paint can be removed using a wire brush after soaking. Do not use the wire brush too harshly. It is easy to damage the bristles on the paint brush. Always brush in the direction of the bristles, working away from the handle.

Steps for Cleaning a Spray Gun

1. **Remove any unused paint from the container.**
2. **Use water to clean water-based paint.** Fill container with water and spray until it emerges clear.
3. **Use white spirit to clean oil-based paint.** Fill container with white spirit instead of water. When clear of paint, clean the container with a mix of hot water and detergent.
4. **Dismantle the machine and clean all parts with a damp cloth.**

Proper Paint Brush and Trash Disposal

Water-thinned (latex) and oil-based paints should be used completely. Be sure to reserve a small amount of the paint for touch-ups. Dry, empty paint containers may be recycled in a recycling program. In most states, latex paint can be dried out in the can and disposed of in your household trash. Leave the lid off to show that the paint has hardened.

Clean Up and Reassemble the Painting Area

No job is complete until the entire area is cleaned up completely and checked for quality. Review all painted areas. Screw all plates back on the wall and check for straightness. Check floors, walls, non-painted areas, and room entrances for paint spatters or drips and clean as needed.

If a client is involved, ask the client to conduct a final assessment of the area with you. Walk through the newly painted areas and point out your work. Note any dissatisfaction and remedy the problem immediately and eagerly. Consider providing the client a small bottle or can of the original paints, clearly labeled, so that they can do minor touch-ups if needed.

Review Questions

1. List three types of paint commonly used in residential and light commercial structures.

2. List the three main types of water-based paint brushes.

3. List the two basic types of oil-based paint brushes.

4. List the steps for preparing a surface for painting.

5. List the steps for fixing a nail pop.

6. List the steps for repairing a small hole in drywall.

Name: _____

Date: _____

Painting Job Sheet 1

Painting

Identify and Select Proper Surface Finish

Upon completion of this job sheet, you should be able to demonstrate your ability to identify and select the proper surface finish and explain the selection process.

Describe the surface being painted:

What needs to be painted:　　bedroom　　bathroom　　living room　　dining room
　　　　　　　　　　　　　　kitchen　　hall　　　　doors　　　　　trim

1 Are surfaces exposed to continual washing or scrubbing? Y/N

If yes, what is the best type of paint to use?

_____
_____

2 Does the surface need to stand up to washing? Y/N

If yes, what is the best type of paint to use?

_____
_____

3 If the room to be painted is a bathroom or kitchen, what is the best type of paint to use?

_____
_____

4 If you are painting a high-traffic area, such as a door or trim, what is the best type of paint to use?

_____
_____

5 If you need to hide minor imperfections, what is the best type of paint to use?

_____
_____

6 If you are painting a ceiling, why is it important to use ceiling paint?

_____
_____

Instructor's Response:

_____
_____
_____
_____

Name: _____

Date: _____

Painting Job Sheet 2

Painting

Select Proper Finishing Tool

Upon completion of this job sheet, you should be able to demonstrate your ability to select the proper finishing tools and explain the selection process.

1. Why is it important to use high-quality tools?
   _____
   _____

2. If it is necessary to remove loose paint from the surface to be painted, what tool(s) should be used to perform this task?
   _____
   _____

3. Paint brushes are selected based on the type of finish being used and the surface being painted. List the two types of brushes and the characteristics of each type.
   _____
   _____

4. Rollers are an alternative to using brushes. Explain the different types of rollers and the types of surfaces they would be used on.
   _____
   _____

Instructor's Response:

_____
_____
_____

Name: _____

Date: _____

Painting Job Sheet 3

Painting

Properly Preparing Surface for Painting

Upon completion of this job sheet, you should be able to inspect the painting surface and determine what preparations need to be done to the surface before painting.

Inspect the following, indicating whether or not inspected surfaces need to be repaired prior to painting.

	Repair	Replace
Walls		
Baseboard		
Crown molding		
Window trim		
Doors		
Door jambs		

If the surface can be repaired, explain what process will be done to repair it. If the surface needs to be replaced, fill out a work order for the work to be done.
Preparation Checklist:

Task Completed

1. Surface repaired or replaced
2. Furniture moved from room or covered
3. Wall plates and receptacles removed
4. Exposed surfaced covered with caulking
5. Drop cloth or other protective material in place

Instructor's Response:

_____
_____
_____

Name: _____

Date: _____

Painting Job Sheet

4

Painting

Applying Paint by Using a Brush, a Roller, or a Sprayer

Upon completion of this job sheet, you should be able to paint the surface using a brush, roller, or sprayer.

1 What does the term "cut an area" during the painting process mean?
_____
_____

2 Is the ceiling being painted? If yes, describe the process that you will use to paint the ceiling.
_____
_____

3 Explain the process for painting the walls.
_____
_____

4 Why is it suggested that you paint in a zigzag or "M" pattern?
_____
_____

5 If using a sprayer, why is it important to pour the paint into a bucket through a strainer?
_____
_____

6 If using a sprayer, why do paint manufacturers recommend thinning the paint first?
_____
_____

Instructor's Response:

_____
_____
_____
_____

Name: _____

Date: _____

Painting Job Sheet 5

Painting

Clean-Up and Storing Equipment

Upon completion of this job sheet, you should be able to demonstrate the ability to clean up and store all equipment.

Task Completed

1. Disassemble paint roller.
2. Clean roller and frame in cleaning solvent.
3. Hang roller to dry.
4. Clean paint brushes in cleaning solvent.
5. Wrap brushes in paper.
6. If using sprayer, remove any unused paint from sprayer.
7. Fill sprayer with appropriate solvent and spray until what emerges is clear.

Instructor's Response:

_____
_____
_____

Chapter 9 Plumbing

OBJECTIVES

By the end of this chapter, you will be able to:

Knowledge-Based

- Identify and select basic plumbing tools for a specific application.
- Identify and select the proper type of plastic piping.
- Identify and select the proper type of copper tubing.
- Identify and select the proper type of metallic pipe.
- Identify and select the proper type of pipe fitting.
- Identify and select pipe hangers and supports.
- Identify, select, and apply caulking.
- Identify and apply caulking.

Skill-Based

- Correctly measure and cut copper tubing.
- Fabricate plastic pipe with correct fittings to correct dimensions as required for job without any leaks.
- Assemble compression fittings without any leaks.
- Clean and replace traps, drains, and vents including use of snake or rod to clean drain lines.
- Caulk and seal fixtures according to manufacturer specifications.
- Fabricate and solder copper pipe with correct fittings as required for job without any leaks.
- Test and set hot water temperature according to manufacturer specifications.
- Follow and apply all national and local building codes.
- Locate and repair leaks in pipes and fixtures.
- Install shower seals.
- Repair, replace, and/or rebuild plumbing fixtures and connections to job specifications without any leaks.

Introduction

Today society relies extensively on the use of plumbing systems to maintain our current standard of living. In a residential setting they are commonly used to transport drinking water to a home while conveying the waste products produced away from the home. Therefore it is essential that the plumbing systems be properly maintained to ensure the health and welfare of the people the system services.

Plumbing Tools

Identify and Select Basic Plumbing Tools

Before any repairs can be made to a plumbing system, some basic tools will be required. These tools can be purchased in most home improvement stores and range in price from a few to a couple hundred dollars. However, like most trades and jobs, quality tools do have a tendency to perform better and last longer.

- Locking tape measure—used to make measurements for plumbing fixtures (sinks, etc.). See Figure 9-1.
- Angled jaw pliers—used to make adjustments to valves and fixtures (Figure 9-2).
- 14-inch pipe wrench—used to loosen and tighten metal pipe and fittings (Figure 9-3).

Note: Typically when tightening fittings on a steel pipe, a second pipe wrench is required: one to hold the pipe and one turn the fitting.

- Hacksaw—used for cutting metal and plastic pipe (Figure 9-4).

Figure 9-1: **Tape measure**

Figure 9-2: **Angled jaw pliers**

Figure 9-3: **Pipe wrench**

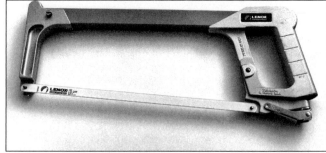

Figure 9-4: **Hacksaw**

Figure 9-5: **Plunger**

Figure 9-6: **Adjustable wrench**

Figure 9-7: **Closet auger**

- Plunger—used to unclog (unstop) a fixture (sink, toilet, etc.). See Figure 9-5.
- 6- and/or 8-inch adjustable wrench—used to loosen and tighten valves and fittings (Figure 9-6).
- Closet auger—used to unstop toilet fixture by removing the foreign object that is causing the problem (Figure 9-7).
- Electric spin drain cleaner—used to unstop kitchen sinks and lavatory. Typically a ¼-inch cable is used for kitchen sinks (Figure 9-8).
- Torpedo level—used to ensure that fixtures are installed correctly (level) and waste piping systems have the necessary slope (Figure 9-9).
- Midget copper cutter—used for cutting copper tubing (Figure 9-10).

Figure 9-8: **Electric spin drain cleaner**

Figure 9-9: **Torpedo level**

Figure 9-10: **Copper tubing cutter**

Figure 9-10A: Plastic pipe cutter

Figure 9-11: Person carrying toolbox (©2006 JupiterImages Corporation)

- Plastic pipe cutter—used for cutting plastic tubing (Figure 9-10A).
- Toolbox—used to store all the plumbing tools (Figure 9-11).
- Miter box—used to cut various precise angles (Figure 9-12).
- Reamer—used to remove burs from piping (Figure 9-13).

Figure 9-12: Miter box

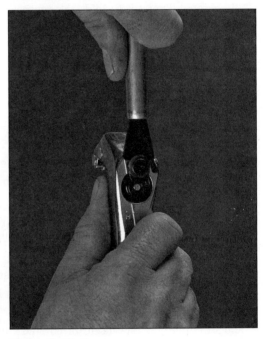

Figure 9-13: Reamer

Piping

Pipe is used to bring potable water and gas into a building and to allow sewage and wastewater to drain from a building.

Types of plastic pipe include:

- PVC (polyvinyl chloride)—available in a variety of sizes and thicknesses (schedules). It is most commonly used in waste water applications. See Figure 9-14.
- CPVC (chlorinated polyvinyl chloride pipe)—a yellowish-white flexible pipe used in water distribution systems. In residential facilities, this pipe is identified as SDR 11 (Figure 9-15).
- ABS pipe (acrylonitrile butadiene styrene pipe)—a black plastic pipe typically used in waste water venting systems (Figure 9-16).

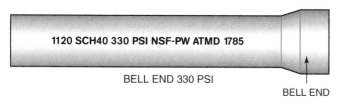

Figure 9-14: **PVC pipe**

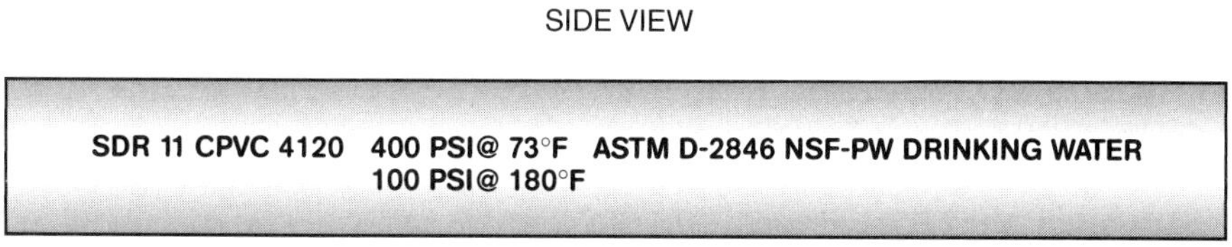

Figure 9-15: **CPVC pipe**

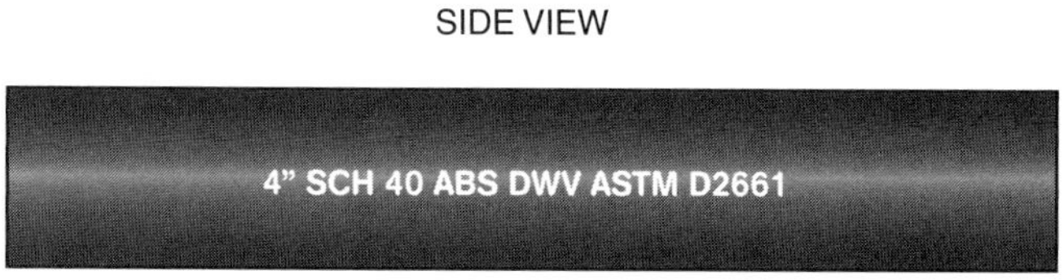

Figure 9-16: **ABS pipe**

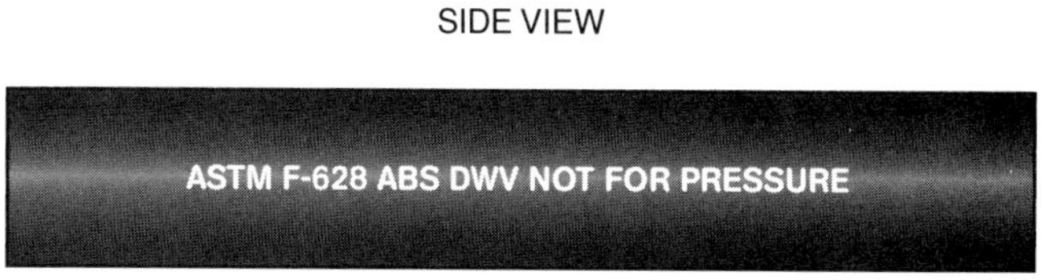

Figure 9-17: **PE pipe**

- PE (polyethylene pipe)—used for exterior water services only. This type of pipe is not well suited for application in which it receives direct sunlight (Figure 9-17).
- PEX (cross-linked polyethylene)—typically a whitish-colored pipe used for water distribution systems (Figure 9-18).

Types of metal pipe include:

- Cast iron pipe—often used in residential waste water systems because it is a much quieter pipe for draining applications than plastic. See Figure 9-19.
- Galvanized and black steel pipe—used for residential gas supply piping (Figure 9-20).

Currently, there are three types of copper tubing used for domestic water. They are type K, L, and M. (see Figure 9-21).

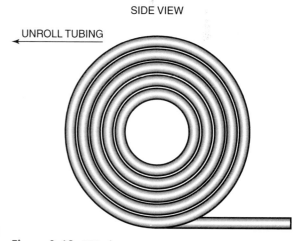

Figure 9-18: **PEX pipe**

- Type K copper tubing—a thick wall tubing used primarily for underground water service. It can be purchased in either soft rolls or in 20-foot stock lengths.
- Type L copper tubing—a thin wall tubing used in above-ground installations. It can be purchased in soft rolls or in 20-foot stock lengths.
- Type M copper tubing—a hard thin wall tubing that is only sold in 20-foot stock lengths. It should be used only in indoor applications in which the tubing can be easily accessed (Figure 9-22 and Figure 9-23).

Figure 9-19: Cast iron pipe

Figure 9-20: Black pipe

Types and Basic Uses of Copper Pipe in a Residential Installation

Copper Type	Potable Water	DWV	Underground	Available in Roll
DWV		✓		
Type M	✓	✓		
Type L	✓	✓	✓	✓
Type K	✓	✓	✓	✓

Figure 9-21: Types and basic uses of copper pipe

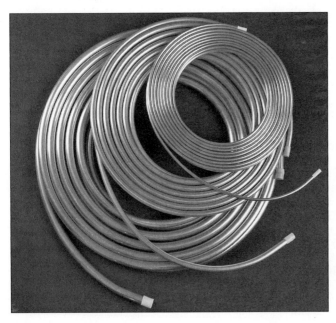

Figure 9-22: Copper tubing

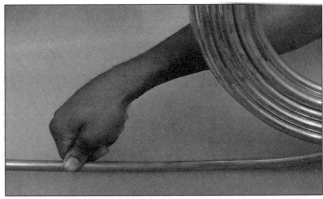

Figure 9-23: Unrolling copper tubing

Pipe Fittings

Regardless of the type of piping materials used in a plumbing system, the basic components (fittings) used to connect the piping together into a useable system are the same (just made from different materials). Some of the more common fittings used (Figure 9-24).

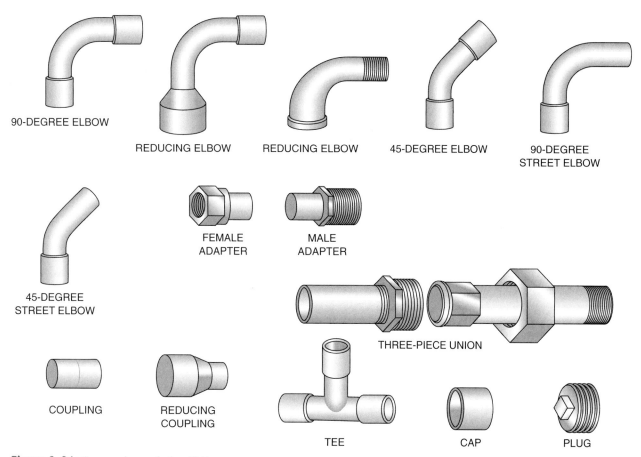

Figure 9-24: **Commonly used pipe fittings**

- 90-degree elbow—used to change the direction of a pipe by 90°.
- 45-degree elbow—used to change the direction of a pipe 45°. It is typically used in pairs to offset a section of pipe around an obstacle.
- Tee—used to create a branch line.
- Cap—used to terminate a section of pipe. A cap is a fitting that fits around the outside of a pipe.
- Coupling—used to connect two pieces of pipe having the same diameter.
- Plug—used to terminate a section of pipe. A plug is a fitting that screws into the inside of a section of pipe.
- Union—used to connect two pieces of pipe together, while allowing for the two pieces of pipe to be disconnected without being cut.
- Reducer—used to reduce a section of pipe to a smaller diameter.

Piping Support and Hangers

Hangers and Supports

In order for a plumbing system to work properly it must be correctly sized and then properly supported. Typically, the type of support used for a particular section of piping depends upon the kind of pipe used, the position of the pipe, and the purpose of that pipe. In general, there are six common types of plumbing support typically used for residential settings. See Figure 9-25 and Figure 9-26.

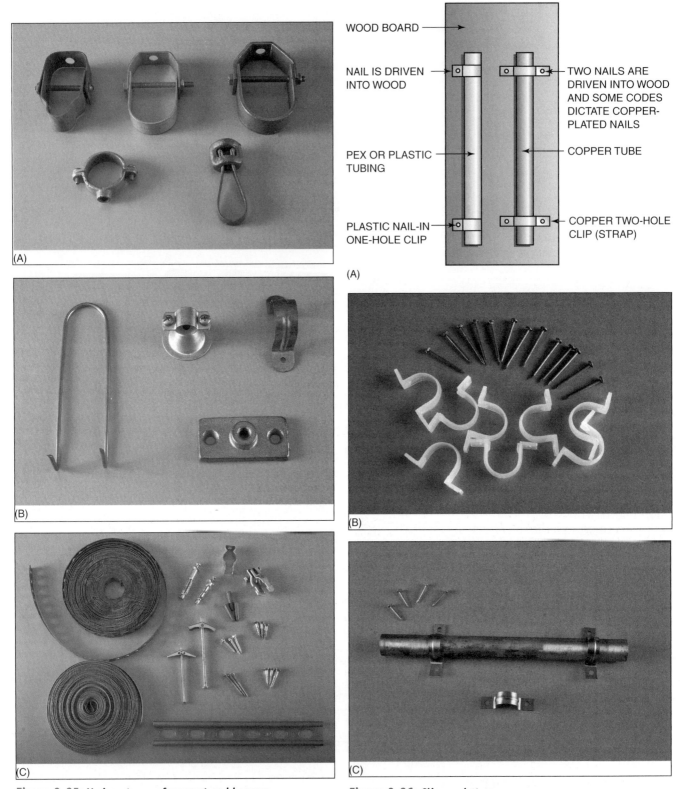

Figure 9-25: Various types of support and hangers

Figure 9-26: Clips and straps

- Clamp—used to anchor a section of pipe to an architectural support.
- Hanger strap—made from a flexible material containing predrilled holes equally spaced throughout the length of the hanger.
- U-bolts—typically used to secure a section of pipe to a structural steel member.

- Wire hanger—Similar to a U-bolt this type of pipe hanger is typically used on smaller lighter pipe.
- Pipe staples—Typically installed using a staple gun, this type of hanger is used to secure piping to an architectural support.

Measuring and Cutting Pipe

To correctly measure copper tubing, the measure is always made from the centers of two fittings or from the end of the pipe run to the center of a fitting (Figure 9-27 and Figure 9-28).

Although copper tubing can be cut using either a hacksaw or a tubing cutter, it is recommended that when possible, a tubing cutter should be used. A tubing cutter will ensure that the end of the copper tubing is square. However, if copper tubing is cut with a hacksaw, it is recommended that a miter box or fixture be used (Figure 9-29).

To cut copper tubing with a hacksaw:

1. Use a locking tape measure to measure copper tubing to the desired length (Figure 9-30).

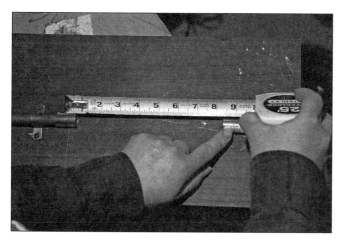

Figure 9-27: **Center-to-center measurement of copper pipe**

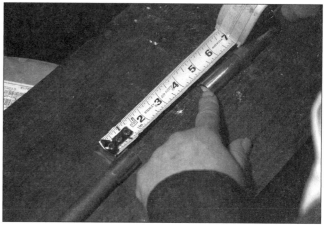

Figure 9-28: **Center-to-end measurement of copper pipe**

Figure 9-29: **Miter box**

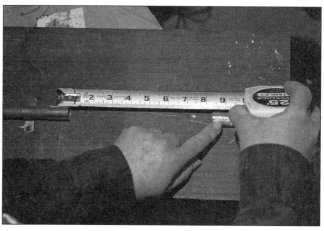

Figure 9-30: **Measuring copper tubing**

2. Place copper tubing into fixture (Figure 9-31).
3. Start the cutting process by pulling and pushing the hacksaw back and forth (Figure 9-32).
4. After tubing has been completely cut; remove any burs inside the tubing with a round file (Figure 9-33).

To cut copper tubing with a tubing cutter, see Figure 9-34.

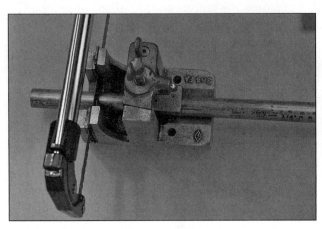

Figure 9-31: **Copper tubing in fixture**

Figure 9-32: **Cutting copper tubing**

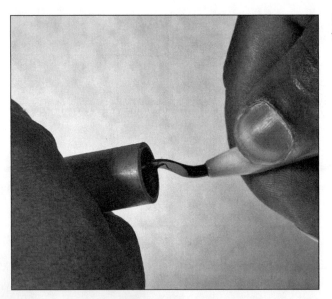

Figure 9-33: **Remove burs from copper tubing**

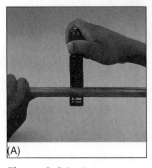
(A)

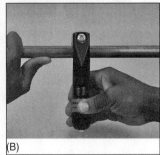
(B)

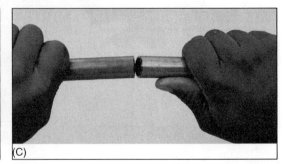
(C)

Figure 9-34: **The proper procedure for using a tubing cutter (A)–(C)**

Unclogging

One of the most frequent types of calls that a facilities maintenance technician will receive is either a clogged lavatory or toilet. These situations usually can be resolved without the assistance of a professional by following a few basic steps.

Using a plunger to unclog a kitchen sink or lavatory:

1. Remove the sink's strainer or plug (Figure 9-35).
2. If the sink does not already have water in it, fill the sink to the halfway mark.
3. Place the plunger's tuber globe over the drain carefully making sure that the entire drain opening is covered by the plunger (Figure 9-36).
4. Using forceful strokes plunge the sink drain at least fifteen times before removing the plunger to see if the sink will drain.
5. If the sink is still clogged, repeat steps 3 and 4.
6. Once the sink has been unclogged, run hot water down the drain for several minutes. This will help clean out anything remaining in the system.

Figure 9-35: **Sink strainer** (©2006 JupiterImages Corporation)

Using liquid drain cleaner to unclog a kitchen sink or lavatory:

1. Before using a liquid drain cleaner, carefully read and follow all instructions located on the package.
2. Care should be taken when handling liquid drain cleaner *not* to spill it on any of the surrounds or on human skin. If you do come in contact with the drain cleaner, consult the packaging or a doctor immediately.

Using a more natural approach to unclog a kitchen sink or lavatory: In some cases, a clog can be removed by using a more natural, environmentally friendly approach. This approach uses baking soda and vinegar. When baking soda is mixed with vinegar, a chemical reaction is produced in which carbon dioxide is released.

1. Remove any contents from the sink.
2. Pour approximately ¼ cup of baking soda into the drain opening.
3. Pour approximately 1 cup of vinegar into the drain opening.
4. Cover the drain opening for fifteen minutes with a lid.
5. Uncover the drain opening and test the drain.
6. If the drain is still clogged, repeat steps 2–5.
7. Rinse the drain with hot water.

Figure 9-36: **Positioning plunger over drain opening**

Note: A plunger may be required to help loosen the clog.

Using a plunger to unclog a toilet:

1. Insert the plunger into the toilet bowl, fully covering the drain opening (Figure 9-37).
2. Pushing down and pulling up on the handle of the plunger, vigorously plunge the toilet fifteen to twenty times.
3. Lift the plunger from the drain opening to see if the toilet drains.
4. If the toilet is still clogged, repeat steps 1–3.

Using a drain cleaner to unclog a toilet:
Note: A liquid drain cleaner should be used on a toilet only as a last resort.

Figure 9-37: **Plunger covering drain opening**

1. Before using a liquid drain cleaner, carefully read and follow all instructions located on the package. Also, before using a liquid drain cleaner on a toilet, make sure that the cleaner is safe to use with porcelain.
2. Care should be taken when handling liquid drain cleaner *not* to spill it on any of the surrounds or on human skin. If you do come in contact with the drain cleaner, consult the packaging or a doctor immediately.
3. After the toilet unclogs, flush the toilet several times to check the flow and remove the drain cleaner.

Cleaning a slow-draining lavatory:

- If a lavatory is slow draining, try running hot water into the sink for about 10 minutes. Running hot water can sometimes loosen contaminates that might be causing the sink to drain slowly.
- If using hot water does not completely open the drain, try using an environmentally safe cleaner (baking soda and vinegar—see the previous section on using a more natural approach.

Unclogging a lavatory with a cleanout:

1. Place a bucket under the sink to catch waste water.
2. Using a wrench, open the cleanout (Figure 9-38).
3. A screwdriver may be used to probe around and pull out the clog (Figure 9-39).
4. Replace the cleanout cover and gasket.
5. Run hot water into the sink for 10 minutes.

Note: If the clog is not removed using a screwdriver, then a sink snake can be used to unclog the drain.

Unclogging a lavatory with a cleanout and a sink snake:

1. Place a bucket under the sink to catch waste water (Figure 9-40).
2. Using a wrench, open the cleanout (Figure 9-41).
3. Carefully push the snake into the cleanout opening, moving the tape back and forth to help it navigate the drain pipe (Figure 9-42).
4. When the snake reaches the clog, twist and push the snake until the clog is removed.
5. Replace the cleanout cover and gasket (Figure 9-43).
6. Run hot water into the sink for 10 minutes.

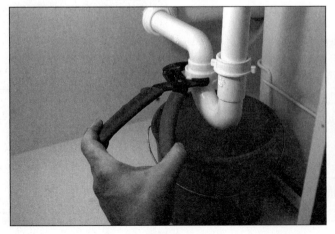

Figure 9-38: Opening a cleanout

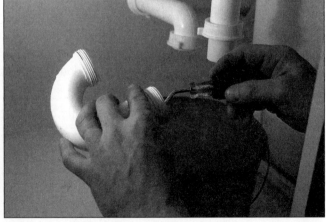

Figure 9-39: **Probing with a screwdriver**

Figure 9-40: **Place a bucket under a sink**

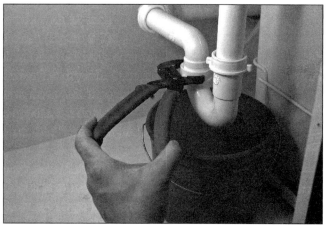

Figure 9-41: **Opening a cleanout**

Figure 9-42: **Using a sink snake**

Figure 9-43: **Replacing a cleanout cover**

Using a plunger to unclog a bathtub drain:

1. Place the plunger's tuber globe over the drain, carefully making sure that the entire drain opening is covered by the plunger (Figure 9-44).
2. Using forceful strokes, plunge the sink drain at least fifteen times before removing the plunger to see if the sink will drain.
3. If the sink is still clogged, repeat steps 3 and 4.
4. Once the sink has been unclogged, run hot water down the drain for several minutes. This will help clean out anything remaining in the system.

Using drain cleaners to unclog a bathtub drain:

1. Before using a liquid drain cleaner, carefully read and follow all instructions located on the package.
2. Care should be taken when handling liquid drain cleaner *not* to spill it on any of the surrounds or on human skin. If you do come in contact with the drain cleaner, consult the packaging or a doctor immediately.

Figure 9-44: **Using plunger in bathtub to unclog drain**

Using environmentally friendly cleaner to unclog a bathtub drain:

1. Pour approximately ¼ cup of baking soda into the drain opening.
2. Pour approximately 1 cup of vinegar into the drain opening.
3. Cover the drain opening for fifteen minutes with a lid.
4. Uncover the drain opening and test the drain.
5. If the drain is still clogged, repeat steps 2–5.
6. Rinse the drain with hot water.

Using a Toilet Auger to Unclog a Toilet

1. Loosen the setscrews on the auger and push the cable into the drain, moving it back and forth, until the clog is reached (Figure 9-45 and Figure 9-46).
2. Tighten the set screws on the toilet auger.
3. While pushing on the toilet auger, crank on the auger clockwise until the obstruction is cleaner.
4. Remove the toilet auger from toilet.
5. Test the flow of the toilet by flushing it several times.

Figure 9-45: Set screws on toilet auger

Figure 9-46: Pushing auger cable into drain opening

Caulking

Caulk is primarily used to seal cracks caused by mating parts. For example, the crack produced by a sink and the mating wall.

Applying Caulk

1. Remove all dust, dirt, and any other grime from around area to apply caulk.
2. If water or a solvent is used to clean the area to be caulked, make sure the area is dry.
3. Insert caulk tube into caulk gun (Figure 9-47).
4. Using a knife, remove the tip of the caulk, cutting away as little as possible. The amount of tip removed will affect the size of the bead of caulk that the caulk gun applies. Make sure the tube of caulk does not contain a second seal. To check, stick a nail into the hole made and cut off the tip (Figure 9-48).

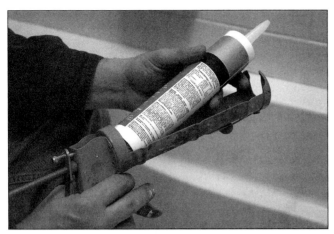

Figure 9-47: Caulk tube and gun

Figure 9-48: Caulk tube with cut angle on end

5. Hold the caulk at about a 5–10° angle, in the direction of travel. The direction of travel is the direction in which you will be pulling the caulk gun (Figure 9-49).
6. After applying the caulk to the crack, use your finger to lightly work the caulk into the crack. (Note: Keeping your finger wet will help ensure a smoother caulk bead. See Figure 9-50.)
7. Use a wet (moist) towel to wipe away all excess caulk.

Figure 9-49: Applying caulking

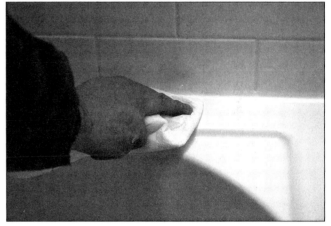

Figure 9-50: Smoothing caulking with your finger

Selecting a Caulk

The four most commonly used caulk today are:
- Acrylic latex caulk—fast drying all-purpose caulk designed for dry applications. It is well suited for painting and is easily cleaned up using soap and water.
- Vinyl latex caulk—designed for wet applications and a good choice for showers and tubs.
- Silicone caulk—long-lasting, mildew resistant, watertight adhesive caulk that does not discolor. Typically, silicone caulk can not be painted or cleaned using soap and water.

- Butyl rubber caulk—designed for use in outdoor applications and can be used to fill large joints or cracks.

Plumber's Putty

Plumber's putty is a pliable sealant used to seal pipe joints and fittings that can be used for quick emergency repairs (Figure 9-51).

Figure 9-51: **Plumber's Putty**

Applying Plumber's Putty

1. Clean the surface of the fitting or fixture where the putty is to be applied.
2. Apply a bead of plumber's putty to the mating part. In the case of a sink drain, apply the putty under the rim of the flange and place the drain into the drain outlet. See Figure 9-52.
3. Tighten the fitting, causing the plumber's putty to spread.

Figure 9-52: **Applying caulk on drain outlet**

Assembling Pipe

Using PVC Cement

See page 243 for a procedure for assembling plastic pipe.

Soldering a Copper Fitting onto a Piece of Copper Pipe

See pages 244–247 for procedures for soldering.
  Tips for safely using a torch:

- *When using a propane torch, always keep a fire extinguisher handy.*
- *Always read and carefully follow all directions on the propane torch.*
- *Always carefully read the instructions on the fire extinguisher before using the propane torch.*

Procedures: Joining Plastic Pipe

- Mark the pipe at the appropriate point with a pencil.

 Cut the pipe using either a hacksaw or tubing shear.

- Remove the burrs from both the inside and the outside of the pipe.

- Apply primer, if required, to both the male and female portions of the joint.

 Apply cement to both the male and female portions of the joint.

C Insert the male end of the fitting into the female end and rotate the pipe ¼ turn.

- Hold the pipe and fitting together for approximately 1 minute to prevent the pipe from pulling out of the fitting.

FROM EXPERIENCE

When working with plastic pipe, always try to dry fit the piping arrangement before cementing. Once a joint is cemented, you don't get a second chance!

CAUTION

CAUTION: Follow all safety guidelines provided on the primer and cement containers. Plastic primers and cements should only be used in well-ventilated areas as the fumes from these chemicals are hazardous to your health.

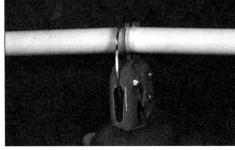

Photo by Bill Johnson.

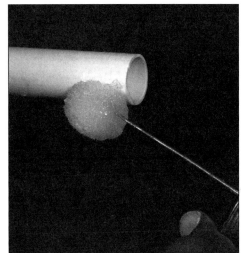

Photo by Bill Johnson.

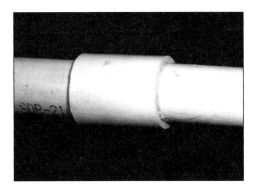

Photo by Bill Johnson.

Procedures: Soldering

- Properly cut and ream the sections to be joined. Refer to the cutting and reaming procedure.

- Using sand cloth or steel wool and the correct size pipe brush, clean the male and female portions of the joint being soldered.

A Using a flux brush, apply flux to the male portion of the joint.

- Insert the male portion of the joint into the female end.

- Before connecting the acetylene regulator to the tank, quickly open and close the stem on the tank using the refrigeration service wrench. This will blow any particulate matter from the opening of the tank.

B Mount the acetylene regulator and torch kit to the tank, making sure that the connections are tight.

C Making certain that the valve on the torch handle is closed, open the stem valve on the acetylene tank ½ to 1 turn using the service wrench. Flip the ratchet on the service wrench so the tank can be closed quickly in the event of an emergency.

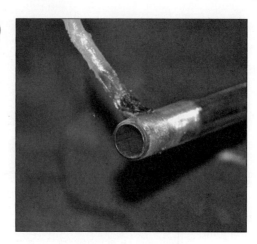

Photo by Bill Johnson.

Photo by Bill Johnson.

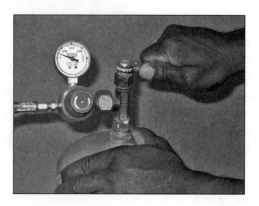

Photo by Bill Johnson.

D Using a soap bubble solution, leak-check the regulator, hose, and torch assembly, making certain that no acetylene is leaking from the kit. Tighten any leaking connections.

E Set the regulator on the tank to the middle range to start. It can always be adjusted later on.

- Open the valve on the torch handle most of the way and ignite the fuel with the striker.

FROM EXPERIENCE

The size of the flame and the amount of heat generated by the torch is directly related to the size of the torch tip used. Larger size pipes will require more heat and will therefore require a larger torch tip. Residential applications typically require the use of an A-3 or A-5 tip for soft soldering.

Procedures — Soldering (Continued)

- Adjust the flame using the regulator on the tank until the desired flame intensity is obtained. A proper flame will have a bright blue inner cone and a lighter blue outer cone. The hottest portion of the torch flame is the tip of the inner cone.

- **F** Apply heat to the joint, placing the tip of the inner cone on the surface of the joint. The flux will begin to melt and flow into the joint. Be sure to keep the flame moving to heat the entire joint. After heating the joint for a short period of time, apply the solder to the joint. The solder should be melted by the heat of the copper, not the heat of the torch flame. If the solder does not immediately begin to flow, remove the solder and heat the joint a little more.

FROM EXPERIENCE

Overheating the joint prior to introducing solder will cause the solder to run off the joint instead of sticking to it. If this should occur, the pipes should be re-cleaned and fluxed to ensure a good solder joint.

CAUTION:
- Always wear safety glasses when soldering or brazing because bubbling flux and liquefied solder can splatter and result in personal injury.
- Never point the torch handle toward the tank or another person while igniting or using the torch!
- Never use a match or cigarette lighter to light the torch. Lighting a match requires the use of two hands and also places your hand very close to the torch flame. Cigarette lighters contain a fuel supply that can easily ignite when the torch is lit.

F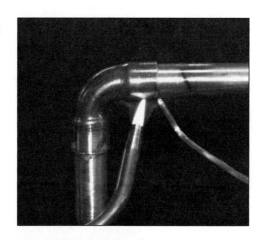

Photo by Bill Johnson.

When the solder begins to flow, feed enough solder into the joint to completely fill it. You should use no more solder than is needed to fill the joint. Using too much solder will result in buildup on the inside of the pipe. Using too little solder will result in leaks.

- To extinguish the torch, simply close the valve on the torch handle. When the torch is no longer needed, close the stem on the acetylene tank and bleed off any acetylene from the hoses by opening the valve on the torch handle.

- While the joint is still hot, it is good practice to wipe the joint with a rag. This removes excess solder and improves the appearance of the solder joint. Applying a small amount of flux to the joint while the pipe is hot also helps clean the joint.

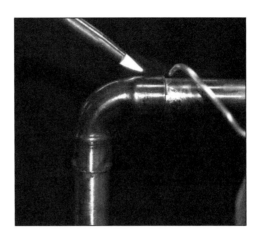

Photo by Bill Johnson.

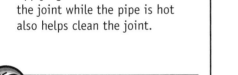

FROM EXPERIENCE

The amount of soft solder used on any given joint should not exceed a length equal to twice the diameter of the pipe.

FROM EXPERIENCE

Molten solder will flow toward the source of heat. To help ensure that the solder flows into the joint, apply heat to the base of the female portion of the joint.

Plumbing Codes

National and Local Plumbing Codes

Because local plumbing codes vary from location to location in the United States, it is recommended that you review all local plumbing codes before attempting any plumbing installation and/or repair. The national plumbing code can be accessed by going to http://emarketing.delmarlearning.com/downloads/StayCurrent_02_05.pdf. Local plumbing codes can be obtained by contacting the state code and administration department.

Adjusting the Temperature of a Water Heater

Recommended Water Temperatures

The U.S. Department of Energy states that a temperature at the tap of 90°F is adequate for most household applications. However, to obtain this temperature the water leaving the tank should be no less than 130°F. This temperature prevents dangerous bacterial growth.

Testing the Water Temperature

1. At one of the sinks in the structure serviced by the hot water tank, turn on the hot water. Be sure to leave the water on for a few moments, making sure the tap water has reached it full temperature.
2. Fill a glass with the hot water.
3. Place a thermometer in the glass of hot water.
4. Remove the thermometer and read the recorded temperature.

Adjusting the Temperature of an Electric Water Heater

The water temperature of an electric water heater is usually controlled by a thermostat located on the side of the water heater. To reset or set the temperature of an electric water heater, simply set the thermostat to the desired temperature.

In some cases, the thermostat may be concealed; if this is the case, do the following:

1. Turn off power to water heater.
2. Using a screwdriver, remove the covering, revealing the thermostat.
3. Adjust the thermostat to the desired temperature.
4. Replace the cover.
5. Turn on electricity.

Note: Some electric water heaters have multiple thermostats. If your water heater has multiple thermostats, *both* thermostats must be set on the same temperature (Figure 9-53).

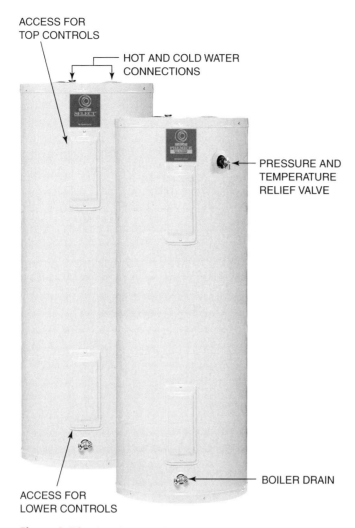

Figure 9-53: Electric water heater

Adjusting the Temperature of a Gas Water Heater

Most gas water heaters have the thermostat located on the side of the water heater; it can be adjusted by simply setting the thermostat to the desired temperature (Figure 9-54). In a few cases, the thermostat may be concealed. If this is the case, then the following steps should be followed:

1. Using a screwdriver, remove the thermostat cover.
2. Adjust the thermostat to the desired temperature.
3. Replace the thermostat cover.

Basic Water Heater Replacement

When you are replacing a water heater, stick with what is already there. If you have an electric water heater, replace it with an electric water heater unless you are willing to run gas lines and exhaust vents. If you are replacing a gas water heater, replace it with a gas water heater unless you are willing and able to install new electrical service.

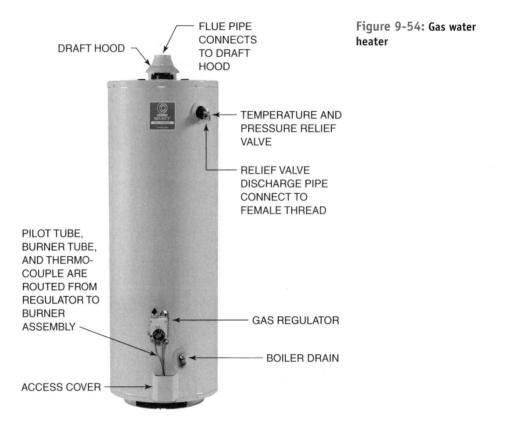

Figure 9-54: Gas water heater

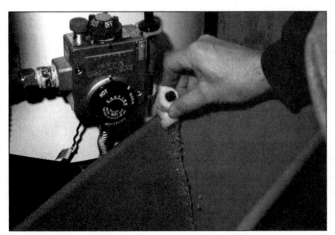

Figure 9-55: Drain the water heater

Figure 9-56: Disconnect water lines

1. Turn off the gas or electricity to the heater.
2. Drain the heater. Opening a hot water faucet will let air into the system (Figure 9-55).
3. Disconnect the water lines (Figure 9-56).
4. Move the new heater to its location by "walking" it or using an appliance cart, dolly, or hand truck. Position the new heater so the piping—particularly a gas vent pipe—will reach most easily.
5. If you removed the shutoff valve, replace it (Figure 9-57).
6. Install the water lines and pressure relief line (Figure 9-58).

Note: Follow manufacturer instructions for a specific step-by-step list on how to install specific water heaters such as gas or electric.

Figure 9-57: Water heater shutoff valve

Figure 9-58: Install water lines

Plumbing Leaks

Plumbing leaks can cause thousands of dollars worth of damage if left unchecked. To check a plumbing system for leaks, the facilities maintenance technician should periodically check under sinks, around toilets, and in crawl spaces. Once a leak has been detected, the problem should be corrected or a plumber should be consulted.

One common source of plumbing leaks is the flapper assembly in a toilet. If the flapper is leaking then the defective flapper should be replaced. A new flapper can be obtained at your local plumbing supplier. To replace the flapper in a toilet, consult the instructions on the flapper package (Figure 9-59).

Figure 9-59: Flapper assembly package

Shower Seals

If the bathroom floor is getting wet after a shower, then possibly the shower door seal should be replaced. Shower seals can be obtained at your local plumbing supply. To replace the shower seal, consult the instructions located on the packaging of the shower seal.

Repairing a Faucet

Sink Faucet

There are several types, makes, and models of sink faucets available on the market today. The most common types of sink faucets available are single and double handled. Faucets manufactured today are either a compression or washerless type.

Compression faucets regulate the flow of water by applying pressure onto a rubber washer located within the faucet, while washerless faucets use a cartridge, ball, or disc to control the flow of water (Figure 9-60).

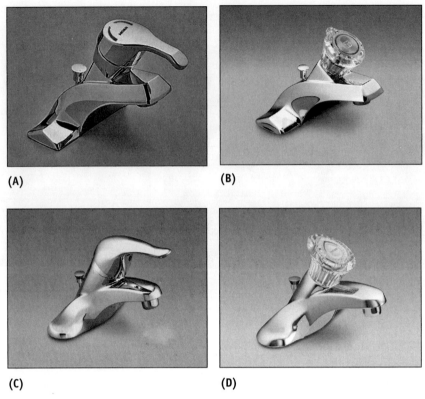

Figure 9-60: **Various faucet types**

Before repairs are made to a faucet, always close the sink drain to prevent parts of the faucet from going down the drain. In addition, never use a pipe wrench on a polished fixture without first applying tape to the pipe wrench jaws. Applying tape to the pipe wrench jaws will prevent the wrench jaws from damaging the fixture's finish. Finally, shut off all water supplies to the faucet.

Because there are a number of types and styles of faucets on the market today, facilities maintenance technicians should consult the manufacturer's documentation at the local plumbing supply when repairing a sink faucet. Only manufacturer-approved parts should be used. In addition, always read and follow the faucet manufacturer's instructions.

Review Questions

1. List three types of plastic pipe commonly used in plumbing.

2. List the two main types of metal pipe used in residential and commercial plumbing.

3. List the three types of copper tubing used in residential and commercial plumbing.

4. List the steps for cutting copper tubing with a hacksaw.

5. List the steps for using a plunger to unclog a kitchen sink.

6. List the steps for unclogging a kitchen sink by using a natural approach.

7. List the steps for unclogging a lavatory with a cleanout.

8. List the steps for using drain cleaner to unclog a bathtub.

Name: _____

Date: _____

Plumbing Tool Identification Job Sheet

Plumbing Tool Identification

Upon completion of this job sheet, you should be able to identify and know the use of any tools used for plumbing tasks at your facilities.

1 Inspect your facility. Identify and list the use of any tools used for plumbing tasks.

Tool	Use

2 Where are the tools stored?

3 Are instructions available for using the tools correctly? If not, do you think they should be?

Instructor's Response:

_____
_____
_____
_____

Name: _____

Date: _____

Plumbing Job Sheet

2

Plumbing

Pipe Identification

Upon completion of this job sheet, you should be able to identify and know the use of pipes used at your facility.

Inspect your facility. Identify and list the uses of various types of pipes.

Pipe Type	Use

Instructor's Response:

_____
_____
_____

Chapter 10: Heating, Ventilation, and Air-Conditioning Systems

OBJECTIVES

By the end of this chapter, you will be able to:

Skill-Based

- Perform general maintenance procedures including:
  - General maintenance on a furnace
  - Tightening and/or replacing belts
  - Adjusting and/or replacing pulleys
  - Replacing filters on HVAC units
- Maintain the heat source on gas-fired furnaces.
- Perform general maintenance of hot water or steam boilers.
- Perform general maintenance of an oil burner and boiler.
- Repair and replace electrical devices, zone valves, and circulator pumps.
- Light a standing pilot.
- Perform general maintenance of a chilled water system.
- Clean coils.
- Lubricate motors.
- Follow systematic diagnostic and troubleshooting practices.
- Maintain and service condensate systems.
- Replace through-the-wall air conditioners.

Introduction

The primary function of heating, ventilation, and air-conditioning (HVAC) systems is to provide healthy and comfortable interior conditions for occupants. This chapter on HVAC will provide a practical description and overview of the various HVAC equipment and systems used in both residential and commercial buildings.

Perform General Furnace Maintenance

There are three common types of furnaces: gas, electric, and oil. Because each furnace manufacturer has a different set of specifications, the following are general guidelines for performing furnace maintenance. For more detailed instructions on maintaining a furnace, read the manufacturer's specifications or call a qualified repair person.

Tightening Belts

Loose belts on an air distribution system can cause the following:

- Insufficient airflow
- Evaporator coil freezing
- Inadequate cooling

If a belt is slipping, the inside surfaces of the pulleys will become polished to a near-mirror finish. If such is the case, the pulleys must be replaced. See the "Replacing Pulleys" section for more on this. If the pulleys are not polished, proceed to adjust the belts:

1. Make certain that the power to the blower is off and that the blower itself has come to a complete stop.

   Safety note: Never attempt to stop rotating equipment or machinery with your hands. Severe personal injury can result.

2. Open the blower access panel/service door on the furnace. Check manufacturer's specifications to locate the blower access panel/service door.
3. With a pencil, mark the position of the motor mounts on the furnace and the bolt positions on the motor base (Figure 10-1).
4. Using the proper sized wrench, loosen the motor mount bolts so that the motor and the motor mount can move freely. Do not completely remove these bolts. Make certain to leave the motor secured to the motor mount itself (Figure 10-2).

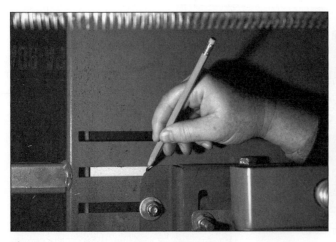

Figure 10-1: **Mark position of motor mounts**

Figure 10-2: **Loosen the motor mounts**

CHAPTER 10 Heating, Ventilation, and Air-Conditioning Systems 261

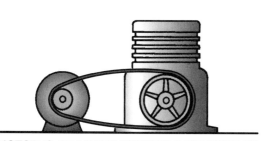

MOTOR IS ADJUSTED TOWARD COMPRESSOR FOR BELTS TO BE INSTALLED.

Figure 10-3: **Adjust the motor to remove the belt**

Figure 10-4: **Technician inspecting the belt**

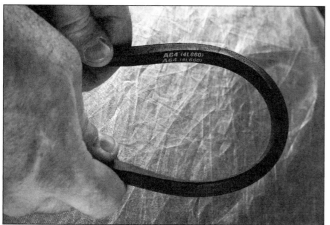

Figure 10-5: **A good belt**

Figure 10-5A: **This belt is cracked and needs to be replaced**

5. Gently move the motor closer to the blower shaft and pulley to further loosen the belt (Figure 10-3).
6. Slide the belt off the pulleys.

Service note: Do not use a screwdriver or other similar object to pry the belt off the pulleys. Doing so can result in the slippage of the tool, causing severe personal injury.

7. Turn the belt inside out and inspect the underside for any cracks, missing pieces of the belt material, or other signs of excessive belt wear. (See Figures 10-4, 10-5, and 10-5A.)
8. If the belt is worn, it needs to be replaced. Refer to the "Replacing Belts" section for more information on this. If the belt is in good condition, go to step 9.
9. With the belt removed, inspect the interior surfaces of the pulleys. The interior surfaces of the pulleys should look rough and should not be shiny. Shiny, polished surfaces on the pulleys are an indication that the belts have been slipping. Belt slippage will cause premature wear. Polished pulleys should be replaced. See the section on "Replacing Pulleys" for more on this (Figure 10-6).
10. If both the belt and pulleys are in good condition, position the motor close to the blower pulley and replace the belt on the pulleys.

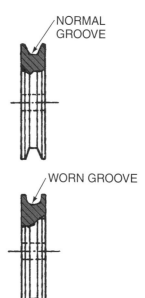

Figure 10-6: **Comparison between normal and worn pulleys**

Figure 10-7: Check the belt tension

Figure 10-8: Tension Gage

Service note: Do not use a screwdriver or other similar object to replace the belt on the pulleys. Doing so can result in the slippage of the tool, causing severe personal injury.

11. Gently push the motor away from the blower pulley to increase the belt tension.
12. When the motor mount is slightly past the pencil markings that were made earlier, begin to tighten the motor mounts to the chassis. Make certain that the new position of the motor mount on the chassis is parallel to the original markings to help ensure proper pulley alignment.
13. Check the belt tension by placing your thumb and fingers on opposite sides of the belt and gently squeezing them together. Basically, you are trying to squeeze the two opposite sides of the belts together. There should be some play in the belts, but no more than one inch of deflection (Figure 10-7).

Service note: Ideally, technicians should use a belt tension gauge to ensure that the belt tension on the belts is correct (Figure 10-8).

14. If the sides of the belts can be pushed in more than one inch, loosen the motor mounting bolts again and repeat steps 11 through 13.

Service note: Do not attempt to adjust the pitch on a variable-pitch pulley to tighten the belt tension. Doing so will change the rotation proportions of the drive assembly and change the speed at which the blower turns.

Replacing Belts

If the belt is broken, damaged, or worn, do not reinstall that belt on the system. Damaged and worn belts should be replaced immediately to help ensure the satisfactory, continued operation of the system.

1. Obtain the belt information from the old belt. If the information on the old belt cannot be read, refer to the "Estimating Belt Sizes" section (Figure 10-9).
2. Obtain a new, exact replacement for the old belt.

Figure 10-9: Belt information

Service tip: Obtain spare belts and keep them on or near the equipment. In the event the belt breaks in the future, you'll have the system back up quickly.

3. If the belt has been removed from the pulleys and is damaged, proceed to step 6. If the belt simply broke, continue with step 4.
4. With a pencil, mark the position of the motor mounts on the furnace and the bolt positions on the motor base.
5. Using the proper size wrench, loosen the motor mount bolts so that the motor and the motor mount can move freely. Do not completely remove these bolts and make certain to leave the motor secured to the motor mount itself.
6. Position the motor close to the blower pulley and replace the belt on the pulleys.
7. Gently push the motor away from the blower pulley to increase the belt tension.
8. When the motor mount is slightly past the pencil markings that were made earlier, begin to tighten the motor mounts to the chassis. Make certain that the new position of the motor mount on the chassis is parallel to the original markings to help ensure proper pulley alignment.
9. Check the belt tension by placing your thumb and fingers on opposite sides of the belt and gently squeezing them together. Basically, you are trying to squeeze the two opposite sides of the belts together. There should be some play in the belts, but no more than one inch of deflection.

Service note: Ideally, technicians should use a belt tension gauge to ensure that the belt tension on the belt is correct (see Figure 10-8).

10. If the sides of the belts can be pushed in more than one inch, loosen the motor mounting bolts again and repeat steps 7 through 9.

Estimating Belt Sizes

There will be times that you will not be able to read the information that is contained on an old belt. This can be due to excessive amounts of dirt, age, or simply the destruction of the belt itself. In order to determine important belt information, follow these steps.

1. Measure the center-to-center distance (in inches) between the motor shaft and the blower shaft (Figure 10-10).
2. Multiply the measurement in step 1 by 2.
3. Measure the diameters of the drive pulley and the driven pulley (Figure 10-11).

Figure 10-10: Measure center to center

Note: If the belt sits deep into one or both pulleys, determine the "effective" diameter of the pulley, which is the diameter of an equivalent pulley if the belt was resting at the outer edge of the pulley.

4. Add the two diameters in step 3 together.
5. Divide the result in step 4 by 2.
6. Multiply the result from step 5 by 3.14.
7. Add the result from step 6 to the result from step 2. This gives you the approximate length of the required belt.
8. Measure the width of the old belt. An "A" belt has a width of $^{17}/_{32}$ inch, while a "B" belt has a width of $^{21}/_{32}$ inch (Figure 10-12).

Figure 10-11: Measure pulley diameters

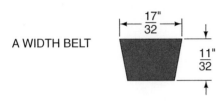

A WIDTH BELT

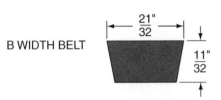

B WIDTH BELT

Figure 10-12: A- and B-width belts

9. The results from steps 7 and 8 provide the type of belt and the approximate length of the required belt.

Sample calculation: Estimate the length of a belt that is used to connect an 8-inch pulley to a 10-inch pulley that is installed on motor and blower shafts that are 16 inches apart.

Here is the step-by-step solution, which corresponds to the steps in the original procedure:

1. The center-to-center distance between the motor shaft and the blower shaft is 16 inches.
2. 16 × 2 = 32 inches
3. Pulley diameters = 8 inches and 10 inches
4. 8 inches + 10 inches = 18 inches
5. 18 inches ÷ 2 = 9 inches
6. 9 inches × 3.14 = 28.26 inches
7. 28.26 inches + 32 inches = 60.26 inches = 60 inches

Adjusting Pulleys

Quite often, the cause for belt slippage, breakage, and premature wear is misaligned pulleys. To check pulley alignment:

1. Make certain that the system is off and all rotating equipment has come to a complete stop. Never attempt to stop rotating equipment by hand. Severe personal injury can result.
2. Remove the access panel on the blower compartment. Check the manufacturer's specifications to locate the blower access panel/service door.
3. Place a straightedge such as a wooden ruler against the faces of both the drive and the driven pulleys (Figure 10-13).
4. The straightedge should touch all four sides of the pulleys:
    a. outside edge of the drive pulley
    b. outside edge of the driven pulley
    c. inside edge of the drive pulley
    d. inside edge of the driven pulley

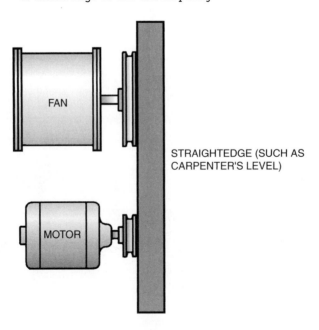

Figure 10-13: Pulleys must be aligned properly

5. This will determine not only if the pulleys are lined up, but also if they are parallel to each other.
6. If the pulleys are not properly aligned but are parallel to each other, either the drive pulley or the driven pulley will have to be repositioned on the respective shaft. See the section on repositioning pulleys.
7. If the pulleys are not parallel to each other, the motor mount or the blower mount must be adjusted to correct this situation.

Replacing Pulleys

If it has been determined that one or more pulleys must be replaced, the original pulley must be removed from the shaft. The shaft may be dirty and/or rusty, making this project potentially very time consuming. Here are some tips to accomplish this.

1. Clean the shaft completely.
2. Spray the shaft with a loosening agent (rust remover). See Figure 10-14.
3. Allow the loosening agent to seep into the joint between the shaft and the hub of the pulley.
4. Completely remove the set screw that holds the pulley to the shaft. This may be a square set screw or an Allen key (Figure 10-15).
5. Be sure to place the set screw in a place where it will not be lost.
6. Spray loosening agent in the set screw hole and allow it to seep into the space between the pulley and the shaft.
7. If the above tips do not help in the pulley removal process, a pulley puller should be used (Figure 10-16).
8. Once the pulley has been removed, clean the shaft completely.
9. Obtain the new pulley and position the new pulley on the shaft so that it is perfectly aligned with the other pulley. Refer to the previous section on aligning pulleys for more on this.
10. Once the pulley has been properly positioned, tighten the pulley securely to the shaft.

Figure 10-14: Spray the pulley hub and shaft

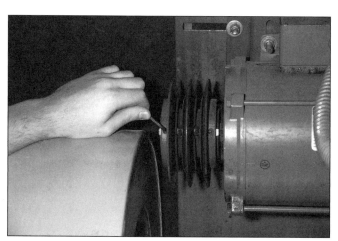

Figure 10-15: Use Allen wrench to loosen set screw

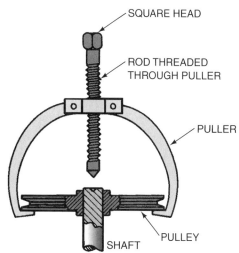

Figure 10-16: A pulley puller

Repositioning Pulleys

There are times when the pulley is in good shape, but needs to be repositioned on the shaft. Use the procedures, steps, and tips in the previous two sections to loosen, reposition, align, and retighten the pulley on the shaft.

Replacing Filters on HVAC Units

The most important thing you can do to keep your air conditioner operating efficiently is to routinely replace or clean its filter(s). Clogged, dirty filters restrict normal airflow, which can cause unfiltered air to carry dirt directly into the evaporator coil. Change air-conditioning filters monthly.

1. Make certain that the system is off and all rotating equipment has come to a complete stop. Never attempt to stop rotating equipment by hand. Severe personal injury can result.
2. Remove existing filter(s) from the system (Figure 10-17).
3. Inspect the channel that holds the filters to be sure that the channels are in good shape and that the filter is supported on at least two sides.
4. Measure the filter channel (Figure 10-18).

Figure 10-17: **Remove the filter**

Figure 10-18: **Measure the filter rack**

5. Make certain that the replacement filter is the same size as the filter channel, not necessarily the size of the filter that came out of the unit. (Someone may have put the wrong size filter in the unit.)
6. Obtain the correct size filter.
7. Locate the arrow on the edge of the filter (Figure 10-19).
8. Install the filter in the channel with the arrow on the filter pointing in the direction of airflow, which is toward the blower.
9. Mark the edge of the filter with the date and your initials (Figure 10-20).
10. Once the filter has been installed, inspect the filter and the channel to be certain that no air is able to bypass the filter.
11. Seal any and all air leaks to prevent/eliminate air bypass.
12. Make certain that any filter channel covers are replaced and secured.

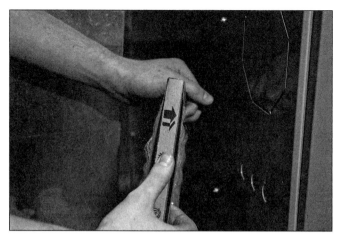

Figure 10-19: Directional arrow on the filter

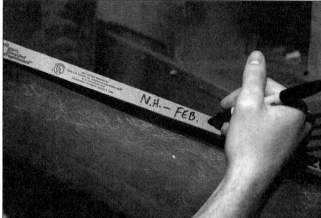

Figure 10-20: Mark or initial the filter

Notes to keep in mind about filter replacement:

- Air that bypasses the filter will cause dirt and dust to accumulate on the coils, blowers, air distribution system components, supply registers, and ultimately end up back in the occupied space.
- Be sure to have an ample supply of filters on hand.
- As an absolute minimum, change filters at the beginning of the heating and cooling seasons. It is recommended, however, to change them every month.
- If the filters are metal permanent-type filters, they can be cleaned.
- If it has been determined that air filters are missing or too small, be sure to visually inspect the return side of the evaporator coil and make certain it is free from dirt and dust.

Maintaining the Heat Source on Gas-Fired Furnaces

Servicing fossil fuel systems should be done by trained professionals, but there are a number of things that the maintenance technician can do to help ensure that the equipment remains in good working order.

1. Typically, the fuel-burning portion of the system does not need to be adjusted each year. So, for best results, do not change any of the current settings on the system.
2. Perform a visual inspection of the burners, pipes, and manifold arrangement. Make certain that these components are free from dirt, dust, and rust. If it is found that there is excessive rust and rust-related damage, call for professional service immediately.
3. If excessive dirt or rust is present, the gas manifold may be carefully removed for cleaning. Follow these steps:
    a. Close the manual gas valve that feeds gas to the appliance.
    b. Disconnect the manifold, making certain to carefully disconnect any components that are attached to the piping (Figure 10-21).
    c. Blow out the manifold with pressurized air, making certain to wear the appropriate personal protection equipment to protect yourself from airborne particles (Figure 10-22).

Figure 10-21: Disconnect the gas manifold

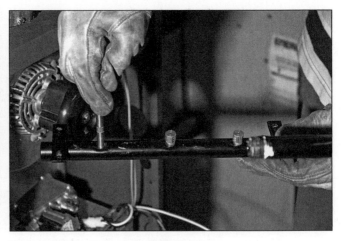
Figure 10-22: Clean the manifold

    d. After the cleaning is complete, reassemble the manifold.
    e. Make certain that all piping connections are tight.
    f. Open the manual gas valve.
    g. Restart the system.
4. Observe the burners when they are lit. The flames should be bright blue with slightly orange tips. If the flames are yellow or are blue with yellow tips, carbon monoxide is present. If yellow flames and/or tips are noticed, seek the assistance of a professional immediately (Figure 10-23a).
5. When burning, the flames should rest just above the burners. Flames that are too high above the burner indicate that there is too much air being introduced. Call for help (Figure 10-23b).
6. Flames should be uniform. Erratic flames may be an indication of a system in need of professional adjustment (Figure 10-23c).
7. Schedule a combustion test/analysis on an annual basis, before the beginning of the heating season.

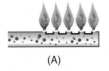 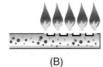 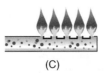
(A)   (B)   (C)

Figure 10-23: Proper and improper flames

Perform General Maintenance of a Hot Water or Steam Boiler

Only experienced contractors should work on boilers. If any of the following conditions are present, a professional should be called in to examine, troubleshoot, and remedy the situation:

1. Water accumulates on the floor around the boiler.
2. Water drips from the pressure relief valve.
3. Heat source fails to energize after one attempt to reset/restart the system has failed.

4. Boiler fails to operate after water has been added to the system.
5. Individual zones fail to heat after attempts to bleed air from the system have failed.
6. Individual zones fail to heat after attempts to troubleshoot zone valves have failed.

Perform General Maintenance of an Oil Burner and Boiler

Oil burners typically require regular service to ensure continued satisfactory system operation. Here is a list of items that must be addressed as well as some suggestions for keeping oil-fired heating systems in tip-top shape.

1. Schedule a combustion analysis before the start of the heating season. If the oil-fired equipment is used to supply domestic hot water year-round, this should be done more frequently. As shown in Figure 10-24, the combustion analysis test should include:
   a. smoke test
   b. carbon monoxide level
   c. carbon dioxide level
   d. stack temperature
   e. draft test

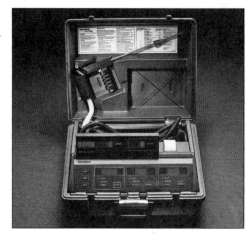

Figure 10-24: **Combustion analysis kit**

2. Keep the oil tank as full as possible. The more oil there is in the tank, the less likely that condensation will form and accumulate in the oil. Water mixed with the oil can result in combustion and operational problems.
3. Clean the heat exchanger on the equipment. To do this follow the manufacturer's recommendations. The steps involve:
   a. Disconnect the flue pipe connection.
   b. Use a boiler brush and vacuum to clean the spaces between the boiler sections.
   c. Wear a protective dust mask (Figure 10-25).
4. Replace the oil filter.
   a. Make certain that the oil valve line is in the closed position.
   b. Unscrew the existing oil filter.
   c. Replace the filter, making certain that the filter canister gasket is in place (Figure 10-26).

Figure 10-25: **Boiler being cleaned**

Figure 10-26: **A gasket**

(A)

(B)

Figure 10-27: Changing the oil filter

Figure 10-28: Cleaning the flue

    d. Make certain that the new filter is tight.
    e. Dispose of the oil filter as you would any other hazardous material (Figure 10-27).
5. Check and clean the flue pipe.
    a. Be sure to wear a protective dust mask.
    b. Make certain that a high quality (filtering) vacuum is used (Figure 10-28).
    c. Be sure to reassemble the flue pipe when finished.
    d. Make certain that the flue pipe is sloped upward toward the chimney.
    e. Inspect the chimney if possible and remove any obstructions or debris from the chimney.
6. Check the oil tank for water accumulation.
7. Inspect the area around the unit for traces of oil.
8. Visually inspect the oil lines for damage.
9. Make certain that fill pipe and vent caps are in place.
10. Check for oil leaks in the tank area.
11. Check for unusual oil odors.
12. Inspect the sight glass (steam boilers only).
    a. If the water level is low, add water to the system via the feed valve.
    b. If the system is losing water at a very fast rate, call for service (Figure 10-29).

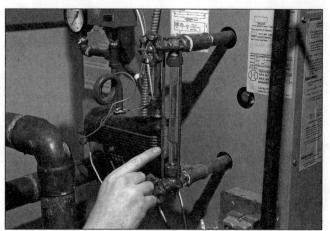

Figure 10-29: Sight glass on a steam boiler

Figure 10-30: **Disconnect the oil line from the burner**

Figure 10-31: **Disconnect the power supply**

13. Remove the burner from the unit.
    a. Place a drop cloth or other barrier between the boiler and the floor.
    b. Make certain that the main oil line is closed.
    c. Disconnect the oil line to the burner, making certain that you have rags on hand for any oil droplets (Figure 10-30).
    d. Disconnect power to the boiler.
    e. Disconnect the wiring connections to the burner, making certain to discharge any capacitors (Figure 10-31).
    f. Place the burner on the floor or work bench.
14. Inspect the combustion chamber.
    a. Look for cracks in the refractory.
    b. Look for and clean up any soot build-up (Figure 10-32).
15. Replace the nozzle (Figure 10-33) and check the firing assembly. You will need to access the manufacturer's guidelines for this, as each manufacturer and each oil burner model has different procedures for accessing/removing the firing assembly.
    a. Make certain that the new nozzle is an exact replacement for the existing part. Be sure to have plenty of spares on hand.

Figure 10-32: **Inside of a combustion chamber**

Figure 10-33: **Replacing the nozzle on the oil heating system**

Figure 10-34: Opening the oil burner

Figure 10-35: Cleaning the cad cell

Figure 10-36: Clean the springs

b. Be careful to not damage the electrode porcelains.
c. Inspect porcelains for damage.
16. Clean the cad cell (on systems that are equipped with them).
    a. Open the top of the oil burner (Figure 10-34).
    b. Locate the cad cell.
    c. Wipe the cell down to remove accumulated dirt (Figure 10-35).
17. Clean the transformer springs.
    a. Double check to make certain there is no power being supplied to the unit.
    b. Open the top of the oil burner.
    c. Clean the transformer springs (Figure 10-36).

Repair and Replace Electrical Devices, Zone Valves, and Circulator Pumps

Here are some general suggestions and tips for replacing electrical devices on heating and air-conditioning equipment.

1. Make certain that *all* electrical power sources are de-energized. Keep in mind that some heating and air-conditioning systems are powered by more than one power source so, even if one source is off, there may still be power to some system components.
2. Make certain that all system capacitors are discharged to avoid receiving an unexpected electric shock.
3. Make certain that an exact replacement, whenever possible, for the component being replaced has been obtained.
4. In the event an exact replacement component is not available, make certain that the replacement component ratings match those of the original as closely as possible.

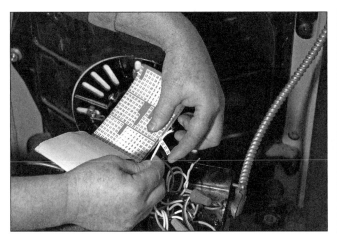

Figure 10-37: **Tag the wires**

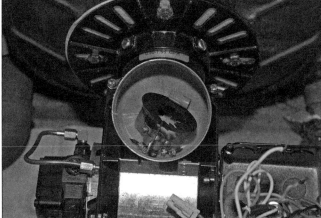

Figure 10-38: **Place screws and small parts in a plastic cup**

5. When replacing components that are directly connected to a water-carrying piping arrangement, make certain that the water from the system has been completely drained to avoid an unexpected flood.
   a. If a zone valve motor is being replaced, there is no need to drain the water system, as the valve mechanism will remain intact.
   b. If the entire zone valve is being replaced, however, the water system must be drained, as the water circuit will be accessed.
   c. If the motor on a circulator pump is being replaced and the linkage/impeller assembly is remaining in the system, the water does not need to be drained.
   d. If the entire circulator is being replaced, the system must be drained.
6. Disconnect the electrical splices one at a time, making certain to label or tag the wires so that you will be able to identify the wires when it comes to reconnecting them (Figure 10-37).
7. Carefully remove the defective component, making certain to place any screws and other small parts in a container such as a cup to prevent them from getting lost (Figure 10-38).
8. Mount the new component in place before making the electrical connections.
9. Once the device has been securely mounted, begin connecting the wires one at a time, making certain that the new wiring corresponds to the wiring of the original device. Be sure to use wire nuts or other mechanical connectors (Figure 10-39).
10. Once completed, make certain that the electrical connections are tight and that no bare wire is extending from the underside of the wire nuts (Figure 10-40).
11. Make certain that no bare current-carrying conductors are making contact with the frame or casing of the device.
12. Make certain that all ground wires are properly connected (Figure 10-41).

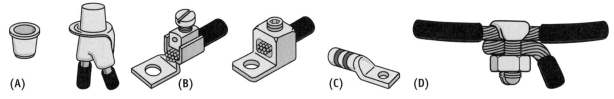

Figure 10-39: **Types of wire connectors**

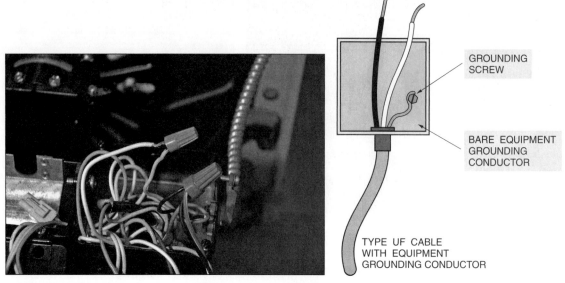

Figure 10-40: Side-by-side nut connections done correctly and incorrectly

Figure 10-41: Ground wire connected to the metal box

Lighting a Standing Pilot

When you perform any procedure on a piece of equipment, it is always recommended that the manufacturer's recommendations and procedures be used before general procedures and suggestions. Here are the steps used to light or relight a standing pilot.

1. If the gas valve is in the ON position, close the valve by turning the knob to the OFF position and allow 15 minutes for any unburned fuel to rise through the appliance.
2. Set the gas valve knob to the PILOT position and push the knob into the valve (Figure 10-42).
3. At the same time, light the pilot by using a large barbeque-type match to avoid coming in close contact with the pilot light (Figure 10-43).

Figure 10-42: Standing gas valve

Figure 10-43: Light the pilot

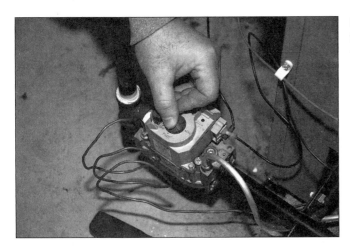

Figure 10-44: **Turn gas valve to ON**

4. Once the pilot light is lit, continue to depress the gas valve knob for about one minute.
5. After one minute, release the knob.
6. The knob should pop up and the pilot should remain lit. If the pilot light goes out, repeat steps 2 through 5.
7. Turn the gas valve knob to the ON position (Figure 10-44).

Perform General Maintenance of a Chilled Water System

Since chilled water systems contain refrigerants, only qualified air-conditioning technicians should access the refrigeration circuits. The EPA requires that all technicians who work on the refrigeration circuits of air-conditioning and refrigeration system be certified under Section 608 of the Clean Air Act. There are, however, a number of items that can be checked by uncertified maintenance personnel.

1. Inspect the piping circuits for signs of leakage, oil, and damage.
2. Inspect the water pumps.
   a. Measure the amperage of the pumps and verify that the amperage is within an acceptable range.
   b. Listen to the pumps and make note of any unusual or abnormal noises and vibrations.
3. Check the system thermometers.
   a. Water being supplied to the chilled water coil should be in the range of 45°F.
   b. Water returning from the chilled water coil should be in the range of 55°F.
4. Measure the temperature difference between the return air temperature and the supply air temperature. This difference could be between 16°F and 20°F.
5. Make certain that all air filters on the air distribution system are clean.
6. Make certain that the blower/motor assembly is operational.
   a. Check pulley alignment.
   b. Check belts for cracks and damage.
   c. Check motor amperage.

Clean Coils

If air filters are properly installed and air is not permitted to bypass the filters, there should be very little, if any, dirt accumulation on the return air side of the evaporator coil. On occasion, though, system air filters are removed or not properly installed and air is permitted to bypass. In order to determine that the evaporator coil is actually clean, visually inspect the coil. This may be a difficult task given the configuration of the system and the installation practices employed. Examining the coil may be as easy as removing the access panel on the air handler or may involve having to cut an access door into the duct system if the coil is mounted on top of a furnace. In any event, once the coil has been inspected and it is determined that the coil is indeed in need of a cleaning, here are some tips and suggestions for doing so.

1. Make certain that the system has been turned off and that all rotating machinery has stopped.
2. Make certain that the area is well ventilated as some cleaning agents give off fumes that may irritate the skin or eyes.
3. Wear proper personal protection equipment such as safety glasses and gloves.
4. Using a brush, remove as much of the dirt as possible (Figure 10-45).
5. Avoid using a rigid wire brush, as the bristles may cause damage to the coil.
6. Avoid flattening the fins on the coil, as this will have a negative effect on the airflow through the coil.
7. Avoid getting cut on the evaporator coil fins. They are sharp and cuts received from them are very painful.
8. Mix the coil cleaner as directed on the product label (Figure 10-46).
9. Apply the water/cleaner mixture to the coil with a high-pressure sprayer and allow it to sit, allowing the chemicals to break up the accumulated dirt on the coil surface. Make certain that the strength of the high-pressure sprayer is weak enough to prevent the bending of the coil fins (Figure 10-47).
10. After the manufacturer's suggested time period, rinse the coil with high-pressure water.
11. Do not use a water hose connected to the building's water supply, as this can damage the coil fins and saturate the duct system.

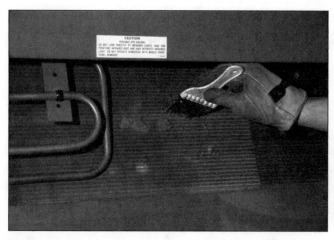

Figure 10-45: Brush the evaporator coil

Figure 10-46: Directions on product label

Figure 10-47: Spray the coil cleaner on the coil

Figure 10-48: Comparison of clean and dirty coil

12. Depending on the amount of dirt on the coil, it may be necessary to repeat steps 9 and 10 to ensure that the coil is as clean as possible.
13. Once the coil has been cleaned, reseal the duct if it was necessary to cut in an access door. If such is the case, be sure to avoid damaging or piercing the refrigerant lines with screws (Figure 10-48).

Lubricate Motors

Periodic motor lubrication is often required to keep the motors in good working order. Permanently lubricated motors do not need to be lubricated, but all others do. Unless otherwise specified, use a medium (20-weight) oil.

1. Remove the oil port plugs from the oil ports on the motor. Typically, there are two oil ports on a motor, so be sure to remove them both (Figure 10-49).
2. Place the oil plugs in a safe place to avoid losing them. Not replacing the oil plugs can allow dirt and dust into the motor, affecting the operation of the component
3. Insert the oil port on the oil container into the oiling tubes (Figure 10-50).

Figure 10-49: Remove oil plugs from motor

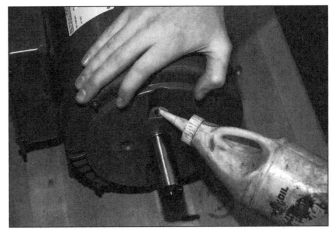
Figure 10-50: Insert oil spout into the oil port

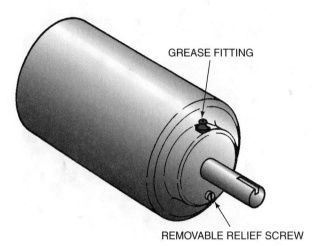

Figure 10-51: Grease fitting

4. Depending on the motor, you should add between 3 and 6 drops of motor oil into each port.
5. Make certain to replace the oil port plugs.
6. Do not over-lubricate motors.
7. If the motor is equipped with grease fittings, the lubrication process is different:
   a. Loosen and remove the relief screw on the motor. The relief screw is located on the opposite side of the motor as the grease fitting (Figure 10-51).
   b. Using a grease gun, pump grease into the motor until grease leaves through the relief screw.
   c. Replace the relief screw.
   d. Repeat this for both sides of the motor.

Follow Systematic Diagnostic and Troubleshooting Practices

When you attempt to diagnose a system problem, using a systematic approach is best. Consider these tips when you encounter a problem.

1. If the system appears to be OFF when it is turned on and operation is desired, check the main power supply to the piece of equipment. Very often, a circuit breaker or switch may have been inadvertently turned off.
2. Check for proper airflow through the air distribution system (furnace and air-conditioning applications).
   a. Reduced airflow can be responsible for a multitude of problems.
   b. Check air filters.
   c. Check belts and pulleys.
3. Check the operational controls on the system.
   a. If a zone valve is not opening, make certain that the thermostat for that zone is calling for heat.
   b. If a boiler is not operating and the water pipes are hot, the water temperature may have reached the desired temperature.
   c. If an air-conditioning system is not operating, make certain that the thermostat is set for cooling before calling for service.
   d. Check for low voltage by switching the fan switch to the ON position (furnace and cooling applications).
   e. If a boiler fails to operate, check safety devices such as high limit controls, pressure controls, and/or low water cutoff switches.
4. Check for safety or trouble lights.
   a. Many newer controls have trouble and/or diagnostic lights.
   b. Keep all manufacturers' literature on hand. This paperwork contains valuable information regarding steps you can take to keep equipment up and running.
   c. Manufacturers' literature contains the trouble codes that often appear on the control's display.
5. Check system operating temperatures and/or pressures.
   a. Determine which parameters are within acceptable ranges.
   b. Determine which parameters are not acceptable.

6. Write down your findings.
    a. Keep logs of your findings.
    b. This may help evaluate future system problems.
    c. Keep records of what was done and when.
    d. This will also help service technicians who come to the job.
7. Narrow your search.
    a. Eliminate items that are definitely operating properly.
    b. Make a list of possible system problems and examine/eliminate them as needed.
8. Ask yourself *why*?
    a. Be sure to fix the cause, not the effect.
    b. Fixing the effect does not fix the underlying problem.

Maintain and Service Condensate Systems

An integral part of the operation of an air-conditioning system is the ability to remove condensate from the structure. Quite often, condensate pumps are used to accomplish this. Depending on the location of the system, condensate pump failure can result in water damage to the structure. Even if there is no condensate pump being used, a gravity-type condensate removal system can cause damage in the event that that line becomes clogged. Here are some tips and suggestions for maintaining condensate removal systems.

1. Inspect the condensate pan under the evaporator for:
    a. signs of rust
    b. signs of damage
    c. dirt, dust, and debris accumulation
    Repair and clean as needed (Figure 10-52).
2. Test condensate lines by pouring a significant amount of water into the drain pan located under the evaporator coil.
    a. Observe the rate of water drainage.
    b. Stop pouring water into the line if water does not drain.

Figure 10-52: **Damaged condensate pan**

Figure 10-53: Pour water into condensate drain pan

Figure 10-54: Condensate pump

    c. Inspect the area around the condensate drain pan for signs of water (Figure 10-53).
3. Inspect the termination point of the line.
    a. If the line terminates outdoors, observe the end of the line before introducing water to the line and again afterwards.
    b. Make certain that the water is actually leaving the structure.
    c. If the line terminates in a condensate pump, make certain that the water is indeed ending up in the pump.
4. If the line is not draining, use pressurized air to blow out the line. Repeat steps 2 and 3 to ensure that the line is now draining.
5. On systems with condensate pumps, make certain to check the operation of the pump.
6. Make certain that the pump remains plugged in or, better yet, have the pump hard-wired to ensure that there is constant power to the pump (Figure 10-54).
7. Test the pump operation by adding water to the pump and inspecting the end of the discharge pipe connected to the outlet of the pump, as in step 3.
8. Inspect the area around the condensate pump for signs of water.

Replace Through-the-Wall Air Conditioners

If it has been determined that a through-the-wall air conditioner needs to be replaced, following are the steps to replace it (Figure 10-55).

1. Remove the existing unit from the sleeve (Figure 10-56).

Service note: Be sure to place a drop cloth or other protective barrier on the floor below the unit to protect the floor from any sharp edges on the unit.

2. Obtain all information from the unit, including the make, model, serial number, voltage rating, amperage rating, and plug type from the unit.
3. Take all unit measurements as well as the internal (daylight opening) measurements of the existing sleeve.
4. Slide the existing unit back into the sleeve.

Figure 10-55: **Through-the-wall air conditioner**

Figure 10-56: **Wall sleeve for air conditioner**

5. With the acquired information, obtain a replacement unit, making certain that *all* measurements and specifications match those of the existing unit.
6. Uncrate and inspect the new unit.
7. Check to make certain that the sizes are correct, the voltage and amperage ratings are the same as the old unit, and the plug is the same.
8. Remove the old unit from the sleeve.
9. Clean and vacuum out the existing sleeve.
10. Slide the new unit into the existing sleeve and secure it according to the manufacturer's installation literature.
11. Make certain that all shipping materials have been removed from the unit and that the air filter is in place before putting the unit into operation.

Review Questions

1. List three common types of furnaces used in residential and commercial environments.

2. List three symptoms of a loose belt on an air distribution system.

3. List the steps for tightening the belt on an air distribution system.

4. List the steps for replacing a belt on an air distribution system.

5. Estimate the length of a belt that is used to connect an 8-inch pulley to a 10-inch pulley that is installed on motor and blower shafts that are 16 inches apart.

6. How often should an air filter be replaced on a HVAC system?

7. List the steps for lubricating a motor.

8. List the steps replacing a through-the-wall air conditioner.

Name: _____

Date: _____

HVAC Job Sheet 1

HVAC

Perform General Inspection on a Furnace

Upon completion of this job sheet, you should be able to perform a general inspection on a furnace system.

Type of furnace: _____
Model: _____
Last Maintenance Date: _____
Maintained by: _____

1. Describe the general running condition of the furnace.
   _____
   _____

2. List four possible fan motor mechanical problems.
   _____
   _____

3. Arrange the following troubleshooting steps in the correct order.

   Test the system operation
   Gather information
   Verify the complaint
   Complete the service call
   Perform the visual inspection
   Isolate the problem
   Correct the problem

4. List the three key indicators that pulleys are aligned properly on the blowers.

Instructor's Response:

_____
_____
_____
_____

Name: _____

Date: _____

HVAC Job Sheet 2

HVAC

Replacing Belts

Upon completion of this job sheet, you should be able to replace belts on a furnace motor.

Type of furnace : _____
Model : _____
Last Maintenance Date: _____
Maintained by: _____

Task Completed

Task:

___ ① Obtain belt information from the old belt.

___ ② Obtain a new, exact replacement for the old belt.

___ ③ With a pencil, mark the position of the motor mounts on the furnace and the bolt position on the motor base.

___ ④ Loosen the motor mounts so the motor and motor mounts can move freely.

___ ⑤ Replace the belt on the pulleys.

___ ⑥ Position the motor, based on the marks in step 3, to increase belt tension and tighten the motor mounts.

___ ⑦ Check the belt tension.

Instructor's Response:

_____
_____
_____
_____

Name: _____

Date: _____

HVAC Job Sheet 3

HVAC

Lighting a Standing Pilot

Upon completion of this job sheet, you should be able to light a pilot light on a gas furnace.

Type of furnace: _____
Model: _____
Last maintenance date: _____
Maintained by: _____

Task Completed

Task:

___ ❶ Take the access cover off the furnace and look for the gas control knob.

___ ❷ Turn the knob to OFF and allow 15 minutes to allow unburned fuel to rise through the furnace.

___ ❸ Turn the knob until the arrow is pointing to the word "Pilot."

___ ❹ Push the knob to start the flow of gas.

___ ❺ Hold a long match or large barbeque-type match up to the pilot to light the pilot.

___ ❻ Once the pilot light is lit, continue to depress the gas valve knob for about 1 minute, then release it.

___ ❼ Turn the knob to the ON position.

Instructor's Response:

_____
_____
_____
_____

Chapter 11: Appliance Repair and Replacement

OBJECTIVES

By the end of this chapter, you will be able to:

Skill-Based

- Replace a gas stove.
- Replace an electric stove.
- Replace a heating element on an electric stove.
- Replace an oven heating element.
- Repair a range hood.
- Replace a range hood.

Introduction

One of the most common duties of a facility maintenance technician is the replacement and repair of gas and electric appliances. Although some appliances require specialty equipment and training to repair; in most cases the facility technician can still perform basic maintenance on these appliances.

Repair or Replace a Gas Stove

Gas Burner Will Not Light

1. Lift the top of the stove.
2. Remove the burner unit by lifting up the back end of the unit and sliding the front end off the gas-supply lines (Figure 11-1).
3. Use a needle or other sharp object to poke into the pilot hole and clean out any debris. Brush the remaining debris away from the tip with a toothbrush. Hold a lit match to the opening to relight the pilot. Lower the lid and turn on your burners to test them (Figure 11-2).

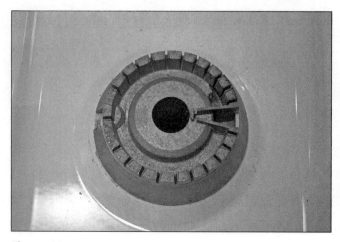

Figure 11-1: Burner unit

Figure 11-2: Cleaning pilot holes on gas burner

Figure 11-3: Spark ignition range on a gas burner

4. Identify a spark ignition range by a little ceramic nub located between two burners. Look for wires running to it (Figure 11-3).
5. Brush away gunk around and on the igniter with an old toothbrush. Clean the metal "ground" above the igniter wire, too. It must be clean to conduct a spark. Close the lid and turn the burner knob to "Light" to test the burner.

The Oven Will Not Heat Properly

1. Remove the oven door (Figure 11-4).
2. Spread a protective covering on the bottom pan to absorb soap and water that may spill during cleaning.
3. Locate the upper burner, attached to the roof of the oven. Using a screwdriver, remove the burner cover, if there is one (Figure 11-5).

Figure 11-4: Removing oven door

Figure 11-5: Upper burner in the oven of a gas stove

4. Using a scrub brush and warm soapy water, clean the burner flame openings to remove the debris. Use a needle or other sharp object to clean out dirt that has collected in the openings (Figure 11-6).
5. Remove the protective covering and lift out the bottom pan. The lower burner will be underneath. Clean the lower burner, following the same procedure described in step 4.
6. Wipe up any water or dirt that collects under the burner. Let all the parts dry thoroughly; then reassemble the oven.

Replace a Gas Stove

1. Shut off the gas line (Figure 11-7).
2. Lay down a piece of plywood to avoid damaging the floor and drag the stove away from the wall. Disconnect the stove from the gas lines.
3. Remove the cover plate at the base of the stove with a screwdriver.
4. If you have old copper tubing, you will first need to disconnect the line at the fitting to the stove. Loosen the nut and slide this back. This is a flared copper type fitting (Figure 11-8).
5. Drag the unit on top of the plywood, avoiding damage to the gas line. Disconnect the old gas line at the coupling. Be careful not to crimp the copper piping (Figure 11-9).
6. If you have natural gas, skip to step 11.

If you have LP gas, you will need to replace the range orifices or spuds (Figure 11-10A and B).

7. Orifices and spuds are individually sized with marks and colors. Loosen the old spud and replace with the new one. The oven orifices will then need to be adjusted.

Figure 11-6: **Cleaning burner flame openings**

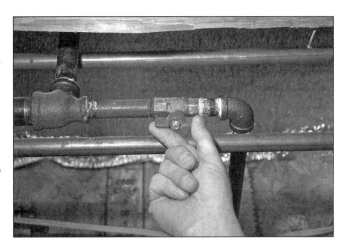

Figure 11-7: **Turn off gas line**

Figure 11-8: **Disconnect copper tubing**

Figure 11-9: **Disconnect gas lines**

(A)

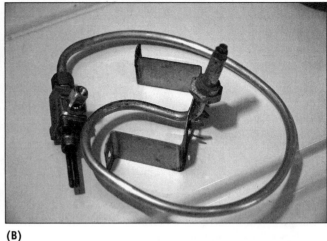
(B)

Figure 11-10: **Range orifices and spuds**

8. Using an appropriate size wrench, tighten down the brass fitting.
9. There may be two orifices, one for the broiler and one for the main oven.
10. Underneath the unit, you will need to reverse the plastic pin in the regulator. Remove the hex nut and flip the plastic pin and reassemble the nut (Figure 11-11).
11. To connect the gas line, use a new connector line with the threads on the pipe identical to the old connector (Figure 11-12).
12. To make the connection, coat the threads with pipe compound.
13. Tighten the fitting and test by applying very soapy water. If it bubbles in a few seconds, you have a leak and will need to fix the leak (Figure 11-13).
14. The anti-tip bracket is mounted to the wall behind the stove per the manufacturer's instructions (Figure 11-14).
15. Move the new stove into place.
16. Level the new stove by adjusting the legs (Figure 11-15).
17. Open the gas line. If there is any smell of gas, shut off the supply and call a qualified service technician.

Figure 11-11: **Regulator underneath the range unit**

Figure 11-12: **Connector line connecting gas lines**

CHAPTER 11 *Appliance Repair and Replacement* **293**

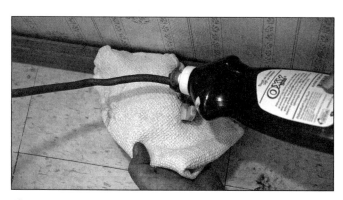

Figure 11-13: **Testing gas line**

Figure 11-14: **Anti-tip bracket**

Figure 11-15: **Level stove**

Repair and Replace Electric Stove

Heating elements eventually burn out. Sometimes, when an element burns out, you can see that the coil burns in two, or blisters and bubbles. When your heating element burns out, you have to replace it because it is not repairable.

Burned-out heating elements are one cause of a burner not working correctly, but it is not the only one. Before you replace the element, troubleshoot and identify the problem.

Troubleshooting the Problem

1. Determine whether the element plugs into a receptacle, as most do, or is wired directly. If it plugs in, move on to step 2. It the element is direct-wired (Figure 11-16), move on to step 4.
2. Remove the plug-in element and inspect the prongs: Lift up the front of the element, then pull the element straight out. Check to see if the prongs are burned, pitted, or otherwise damaged. If they are, you'll need to replace the element and the receptacle (Figure 11-17).
3. If the prongs are clean, test the element: First reinstall it in the receptacle and turn on the burner—sometimes an element just needs to be reseated to work right. If it

Figure 11-16: **Heating elements**

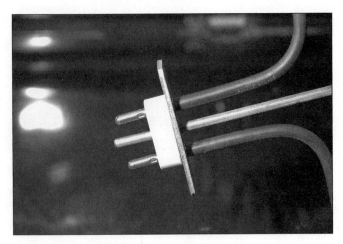

Figure 11-17: **Prongs on heating element**

Figure 11-18: **Porcelain insulators**

still does not heat, turn off the burner, exchange the element with another of the same size, and test. If the burner works now, the original element needs to be replaced.

4. If the element is direct-wired, lift the front of the element and pull it out until you see a white porcelain insulator with clips on each side (Figure 11-18).
5. Use a flat-head screwdriver to open the insulator and remove the clips. Then separate the two halves of the insulator.
6. Remove the screws that hold the element to its wiring. Exchange the element for another of the same size. Reassemble both elements so no bare wires are left exposed, and then turn on the burner. If the new element works, the original element needs to be replaced.

Replacing an Element

1. Get a new heating element identical to the one you are replacing (Figure 11-19).
2. Install the new element in the stove. For a plug-in element, just plug it into the receptacle. For a direct-wired element, screw the new element to its wiring, reassemble the two halves of the porcelain insulator, and snap the clips in place (Figure 11-20).
3. Test the element to make sure it's operating.

Figure 11-19: **Heating elements** (©2006 JupiterImages Corporation)

Figure 11-20: **Installing new heating elements**

Replacing a Receptacle

1. Disconnect the old receptacle. If it is screwed to the cooktop, use a screwdriver to disconnect it. If it is held in place by a spring steel clamp, spread the clamp and pull out the receptacle (Figure 11-21).
2. Lift the cooktop so you can access the receptacle wiring (Figure 11-22).
3. Remove the receptacle. Wrap the wires with masking tape and label them so you can install the new receptacle correctly, then cut the wires (Figure 11-23).
4. Install the new receptacle. Strip the ends of the wires with a wire stripper, then twist the wires together and twist on wire nuts to hold them together. Reinstall the receptacle in the cooktop and install the element (Figure 11-24A and B).

Figure 11-21: **Disconnecting old receptacle**

Figure 11-22: **Cooktop lifted to access receptacle wiring**

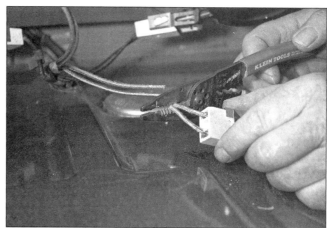
Figure 11-23: **Wires being cut**

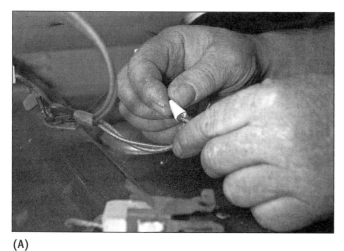
(A)

(B)

Figure 11-24: **New receptacle installed**

Oven Does Not Heat Properly

Test the Thermostat

Figure 11-25: Oven thermometer

1. Place an oven thermometer inside the oven and shut the door (Figure 11-25).
2. Turn on the oven, set it for 350°F and let it heat for 30 minutes.
3. Check the thermometer. Most thermostats are accurate to within 25°F. If the thermostat is off by more than 50°F, the thermostat is bad and you will need to have a professional replace it. If the thermostat is off by less than 50°F, adjust the thermostat.
4. Locate the adjustment screw. On some thermostats, the adjustment screw is on the back of the thermostat knob; on others it is inside the thermostat shaft (Figure 11-26).
5. To make a temperature adjustment on the back of a knob, remove the knob and loosen the retaining screws on the back. Turn the center disk toward "Hotter" or "Raise" to increase the temperature, or toward "Cooler" or "Lower" to decrease the temperature. Tighten the screws, reinstall the knob, and test the oven. Readjust the knob if necessary (Figure 11-26A).
6. To make a temperature adjustment inside the shaft, remove the knob and slip a thin flat-head screwdriver into the knob until it engages the adjustment screw in the bottom. Turn the screwdriver clockwise to raise the temperature, counterclockwise to lower it. Each quarter-turn will move the temperature about 25°F.
7. Reinstall the knob and test the oven. Readjust the temperature if necessary.

Figure 11-26: Thermostat and screw

Figure 11-26A: Thermostat knob

Replace the Oven Heating Element

1. Remove the oven racks so you have access to the element (Figure 11-27A and B).
2. Remove the two screws from the element mounting plate, which sits flush against the back wall of the oven.
3. Pull the element gently out as far as the wire will allow.

(A)

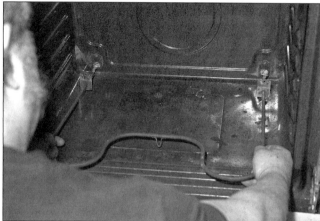
(B)

Figure 11-27: **Oven heating element**

4. Remove the supply wires from the element terminals (Figure 11-28).
5. Replace the element with a new identical one.
6. Put the new element in place and reconnect the leads. There are usually only two wires going to the element; it doesn't matter which wire attaches to which terminal, as long as they're screwed on tight.
7. Push excess wire back behind the insulation.
8. Line up holes and re-install the mounting bracket using the same screws you removed earlier (Figure 11-29).
9. Replace the oven racks.

Figure 11-28: **Element terminals**

Figure 11-29: **Mounting brackets**

Replace Electric Stove

1. Lay down a piece of plywood to prevent damaging the floor and drag the stove away from the wall.
2. Unplug the stove from the electrical outlet.
3. Remove the old stove and replace it with the new one.
4. Plug in the new stove.
5. Move the new stove into place.
6. Level the new stove by adjusting the legs.

Troubleshooting an Ice Maker in a Refrigerator

If the ice maker does not make ice but you can see the arm swing into motion and you hear a buzz for about ten seconds after it is finished, this normally means that there is a problem with the water supply line.

Figure 11-30: **Fill tubes**

1. Check to make sure the water supply line is not kinked behind or beneath the refrigerator. If the ice maker has frozen up, it will need to be unthawed.
2. Unplug the refrigerator.
3. Remove the ice bin and remove loose ice from the ice maker.
4. Find the fill tube, the white rubber-like hose, that delivers the water into the ice maker and pull the small metal cap off of the housing that holds the full tube down (Figure 11-30).
5. Warm the hose and surrounding mechanism to melt any ice blocking the mechanism. This can be done using a hair dryer or soaking the supply tubing in hot water.

Note: Be careful not to melt the plastic parts.

Replacing the Ice Maker

Follow the manufacturer's instructions on replacing the ice maker.

Repairing a Refrigerator

Major problems will require a trained refrigeration technician. However, many times the problem is simple and can be corrected.

Adjust Controls

Refer to the specific refrigerator manufacturer's instructions.

Test and Replace Door Gaskets

1. Test the door seal in several places by closing a piece of paper in the door, and then pulling it out. There should be some resistance, indicating that the door is sealed.
2. Remove the old gasket one section at a time. Some gaskets are held on by retaining strips, others by screws or even adhesive.
3. Install an identical gasket by using the retaining strips or screws, or new adhesive.

Troubleshooting Dishwasher Problems

Note: Before doing any work on your dishwasher, turn off the power at the circuit-breaker box.

Water on the Floor Around the Dishwasher

1. Check your gasket for cracks or deterioration (Figure 11-31).
2. If the gasket is damaged, remove it by unscrewing it or prying it out with a screwdriver. Replace it with the same type of gasket as what was removed. Before installing the new gasket, soak it in hot water to make it more flexible.

Note: A dishwasher can also leak if it is not level.

Clogged Sprayer

1. Remove the sprayer and soak it in warm white vinegar for a few hours to loosen mineral deposits. Then clean out each spray hole with a pointed device such as a needle, awl, or pipe cleaner (Figure 11-32).

Dishwasher Overflows

1. Open the dishwasher door and locate the float switch. It should be a cylinder-shaped piece of plastic and may be set to one side along the front of the cabinet or near the sprayer head in the middle of the machine (Figure 11-33).
2. Check the float to make sure it moves freely up and down on its shaft. (You may have to unscrew and remove a protective cap to get to the float.) If the float sticks, you'll need to clean away any debris or mineral deposits that are causing it to jam.
3. Pull the float off the shaft and then clean the inside of the float with a bottle brush. Clean the shaft with a scrub brush.
4. Reinstall the float and check that it moves smoothly.
5. Set the dishwasher to fill, and check to see if it overflows.

Replace Dishwasher

1. Turn off the power to the dishwasher circuit at the electrical service panel.
2. Shut off the hot water supply to the unit. This is typically under the sink if the supply comes from there, but it may also be under the dishwasher or in the basement if the supply comes through the wall or floor of the dishwasher opening.

Figure 11-31: Gasket around dishwasher

Figure 11-32: Dishwasher sprayer

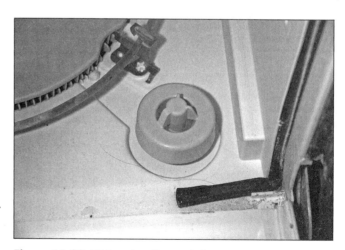

Figure 11-33: Dishwasher float switch

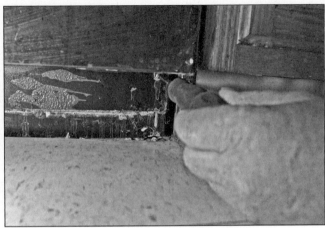

Figure 11-34: **Accessing lower panels**

Figure 11-35: **Electrical box on dishwasher**

3. Remove the access and lower panels at the base of both dishwashers (Figure 11-34).
4. Remove the electrical box from the old dishwasher (Figure 11-35).
5. Unscrew the wire nuts and pull apart the wires. Start with the green wires, then white, and then black.
6. Disconnect the drain hose from the waste tee on the drain line, or the inlet on a disposer, by using pliers to open a spring clamp or a screwdriver to open a screw-type clamp. Do the same where it connects to the dishwasher. If you cannot easily access that connection, you can disconnect it later (Figure 11-36).
7. Disconnect the water supply line from the water inlet on the dishwasher (Figure 11-37).

Note: Have an old towel or drip pan handy to mop up or catch water that will spill out of drain and water lines as they are disconnected.

8. Lay down a piece of plywood to drag the old dishwasher onto.
9. Open the door to access and remove the screws that secure the dishwasher to the underside of the countertop. Then adjust the front leg levelers to lower the unit so you can slide it out (Figure 11-38).
10. Take the new dishwasher out of the box and check the back to verify that all of the connections are there.

Figure 11-36: **Drain hose**

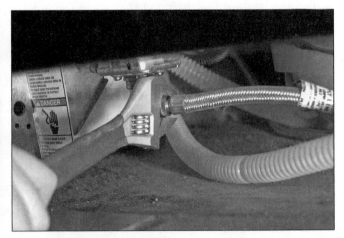

Figure 11-37: **Water supply line**

Figure 11-38: Removing screws in counter

Figure 11-39: Drain line connection

11. Take the cap off the drain line connection at the dishwasher (Figure 11-39).
12. Attach the drain line to the dishwasher.
13. Using pliers, crimp the clamp around the hose to secure.
14. Close and lock the door; then slide/roll the machine to the opening.
15. Adjust the leveling legs as indicated by the manufacturer to raise the dishwasher (Figure 11-40). Use a level to verify that the unit is level. Install the mounting screws into the underside of the counter.
16. Reverse the removal procedures to connect water, drain, and electric lines. Cut off the exposed ends of electrical wires and use wire strippers to strip about ½ inch of insulation from the ends. Twist wires together (white-to-white, black-to-black) and twist on new wire connectors. Secure the ground (green) wire. Tighten the strain-relief connector.
17. Install decorative panels.
18. Follow the manufacturer's instructions to adjust the door so there is an even space on both sides. You might, for example, need to move the door spring to a new mounting hole.
19. Open the water valve to check for leaks in the water line at valve and inlet connections. Restore power and operate the machine to check for drain leaks.
20. Reinstall the lower panel and front access panel.

Figure 11-40: Leveling legs of dishwasher

Repairing a Range Hood

A range hood that does not adequately remove smoke and smells from your kitchen is usually caused by one of the following:

- The grease filter or some part of the exhaust ductwork may be clogged.
- The fan may be bad.

Unclogging the Exhaust Fan

1. Remove the filter and soak in a degreasing solution until the grease is dissolved (Figure 11-41).

(A)

(B)

Figure 11-41: Exhaust fan filter

Figure 11-42: Remove exhaust fan

2. Wash with warm, soapy water to remove any traces of the degreaser. Also, a filter may be put it in the upper rack of the dishwasher and run it through a normal cycle.
3. Remove the exhaust fan. Unplug the fan and remove it from the hood (Figure 11-42).
4. Clean the fan blades with an old toothbrush dipped into a cleaning solution.
5. Clean the inside of the exhaust ductwork, using a plumber's snake with a heavy rag tied around the end. Push the snake through the ductwork. Soak the rag in a cleaning solution and run it through the ductwork. Rinse out the rag and repeat the operation until the duct appears to be clean (Figure 11-43).
6. Clean the exhaust hood that is attached to the outside of your house (Figure 11-44).
7. Reinstall the grease filter.

Figure 11-43: Clean exhaust ductwork

Figure 11-44: Exhaust hood on outside of house

Replacing a Range Hood

1. Remove the old range hood (Figure 11-45).
2. On the new range hood, remove the filter, fan, and electrical housing cover. Remove the knockouts for the electrical cable and the duct (Figure 11-46).
3. Protect the surface of the cooktop with heavy cardboard and set the range hood on top of it. Then connect the house wiring to the hood. Connect the house black wire to the hood black wire and the house white wire to the hood white wire. Then connect the house ground wire under the ground screw and tighten the cable clamp onto the house wiring.
4. Using the mounting screws to install the hood, slide the hood towards the wall until the mounting screws are engaged. Tighten the screws securely with a long-handled screwdriver. Then replace the bottom cover.
5. Fasten the ductwork to the hood using duct tape to secure joints and make them airtight (Figure 11-47).
6. Install the light bulbs and replace the filters. Turn on the power at the service panel and check for proper operation.

Figure 11-45: **Range hood**

Figure 11-46: **Range hood with knockouts**

Figure 11-47: **Hood ductwork**

Repairing Microwaves

Because microwave leakage can be hazardous and high wattage is present, limit your microwave repairs to light bulb changes, if the light bulb is easily accessible, and checking to make sure the oven is getting power. For other repairs, call a qualified technician to make repairs on a microwave oven.

Troubleshoot Washers

If the washer is not getting either hot or cold water, refer to your owner's manual to ensure that the washer is not operating as it should. If the washer isn't operating as it should, there may be a problem with the water inlet valve.

Check the Water Inlet Valve

1. Disconnect the appliance's power supply.
2. Locate the washer's water inlet valve. It will be at the back of the washer, and it will have water hoses hooked up to the back of it (Figure 11-48).
3. Shut off the supply of water to your washer (Figure 11-49).

Figure 11-48: **Inlet valve**

Figure 11-49: **Shutting off water supply**

Figure 11-50: **Disconnect hoses**

4. Disconnect both hoses at the back of the washer. Point the hoses into a bucket or a sink, and then turn on the water supply again. Do this to confirm that you are receiving adequate water pressure, and that there is not some sort of blockage in the line (Figure 11-50).
5. Inspect the screens found inside the valve. Clean out any debris you find. You should be able to pop them out with a flat-head screwdriver. Do use caution when handling the screens as they are irreplaceable.

Replacing Washer Inlet Valves

1. Disconnect the appliance's power supply.
2. Locate the washer's water inlet valve. It will be at the back of the washer, and it will have water hoses hooked up to the back of it (Figure 11-51).
3. Shut off the water supply to your washer.
4. Disconnect both hoses at the back of the washer. Point the hoses into a bucket or a sink, then turn on the water supply again. Do this to confirm that you are receiving adequate water pressure, and that there is not some sort of blockage in the line.
5. Remove the screws that hold the inlet valve in place (Figure 11-52).
6. Remove the hose connecting the valve to the fill spout.
7. Connect your new water inlet valve to the water line.
8. Attach both water supply hoses. Turn the water on and check for leaks.
9. Reconnect the washer to the power supply.

Figure 11-51: **Inlet valve**

Figure 11-52: **Inlet valve**

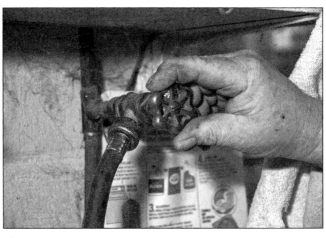

Figure 11-53: **Water faucets**

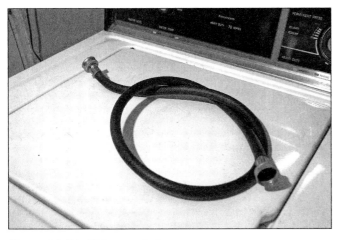

Figure 11-54: **Water-supply hose**

Figure 11-55: **Intake screen**

Washing Machine Fills Slowly

When a washing machine fills slowly, the problem is usually a clogged intake screen.

1. Turn off the water faucets that feed the machine (Figure 11-53).
2. Unplug the washer and pull it far enough away from the wall that you can get behind it to work.
3. Remove each water-supply hose (Figure 11-54).
4. Locate the screens and gently pry out the screens using a small flat-head screwdriver (Figure 11-55).
5. Clean the screens with an old toothbrush.
6. Reinstall the screens.
7. Reinstall the hoses and tighten the couplings securely.
8. Turn on the water to check for leaks; then plug in the machine and push it back into position.

Install a New Washer

1. Turn off power to the old washer.
2. Remove the old washer.
3. Clean and dry the floor and move the new washer into place to be connected.

Figure 11-56: Fasten drain hose

Figure 11-57: Attach hose to the washer

4. Fasten the drain hose to the washer with a hose clamp. Be sure not to tighten it too much or you might strip the screw (Figure 11-56).
5. Attach the water hoses to the washer. The hot and cold on the taps and on the washer are usually clearly marked. Red indicates hot; blue indicates cold (Figure 11-57).
6. Plug the washing machine in and move it into place, placing the drain hose in the drainpipe when you can reach it.
7. Push the washer the rest of the way into the space, being careful not to crimp the hoses.
8. Leave about an inch and a half of space around the washer to allow room for it to vibrate.
9. Turn the water faucets on.
10. Turn the power back on.
11. Run a cycle without clothes or detergent before you use the machine to clear the water pipes and make sure the drainage is adequate.

Troubleshoot a Dryer

Dryer Takes a Long Time to Dry Clothes

The heating element is partially or completely burned out. Follow the manufacturer's instructions for the specific dryer brand to test and replace the heating if necessary.

The Vent Is Clogged

If the dryer feels really hot, but the clothes take forever to dry, a clogged vent could be the problem.

1. Check the vent flap or hood on the outside of the house. Make sure that you feel a strong flow of air coming out when the dryer is running. If not, try cleaning out the vent with a straightened clothes hanger (Figure 11-58).
2. If the vent flap is not the problem, check for a kink or sag in the duct and straighten the hose if necessary.

Figure 11-58: **Vent flap**

Figure 11-59: **Dryer duct**

3. If a kinked or sagging duct is not the problem, disconnect the duct from the dryer and look for blockage inside with a flashlight. To remove the blockage, shake it out or run a wadded cloth through the duct. If the duct is damaged, replace it (Figure 11-59).

Review Questions

1. List the steps for unclogging a vent.

2. What should the technician check if a dryer does not dry clothes?

3. List the steps for installing a washer.

4. List the steps for replacing a washer inlet valve.

5. List the steps for replacing a range hood.

6. List the steps for unclogging an exhaust fan.

7. List the steps for replacing a dishwasher.

8. List the steps for replacing a gas stove.

Name: _____

Date: _____

Appliance Repair and Replacement Job Sheet

1

Replacing an Electric Stove

Upon completion of this job sheet, you should be able to replace an electric stove.

In the space provided below, list the steps for replacing an electric stove.

Instructor's Response:

_____
_____
_____
_____

Name: _____

Date: _____

Appliance Repair and Replacement Job Sheet

2

Replacing a Dishwasher

Upon completion of this job sheet, you should be able to replace a Dishwasher.

In the space provided below, list the steps for replacing a Dishwasher.

Instructor's Response:

_____
_____
_____
_____

Chapter 12 Trash Compactors

OBJECTIVES

By the end of this chapter, you will be able to:

Skill-Based
- Perform general maintenance procedures.
- Perform general maintenance of hydraulic devices.
- Perform a test of the interlock safety device.
- Check the general condition of a Dumpster.

Introduction

Waste and trash are usually collected in a trash can, which is then put out for trash collection. A trash compactor does exactly what its name implies. Instead of putting your trash into a trash can, you put it in the compactor, where it gets compressed to between $1/10$ and $1/12$ the space it would normally take up.

General Maintenance

A trash compactor is a relatively simple appliance with only a few components. Those components include a motor, drive screws, compression ram, limit switches, door switch, exterior controls, and in some units, an odor control system (Figure 12-1).

Common Trash Compactor Problems and Solutions

Compactor will not start:
- Make sure that the door is completely closed.
- Make sure the compactor is plugged in securely.
- Check for a blown fuse or tripped circuit breaker.
- Inspect the electrical cord for damage.

Figure 12-1: **Trash compactor**

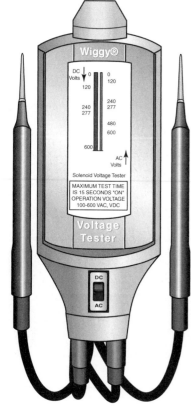

Figure 12-2: **Voltage tester**

The motor runs but trash is not compacted.

- Drawer must be about ⅓ full before any compaction will take place.
- Check the outlet voltage. To measure the voltage at an electrical outlet requires the use of a voltage tester (Figure 12-2).
- Check the power nuts for wear or obstructions. See manufacturer's specifications.
- Check the power screws for wear or obstructions. See manufacturer's specifications.
- Lubricate the power screws. See manufacturer's specifications.
- Check the drive belt/chain/gears. See manufacturer's specifications.

Compactor starts but does not complete cycle (ram is stuck).

- Object in trash may be causing door to trigger tilt switch.
- Check for loose connections.

The drawer is stiff or difficult to open.

- Clean the drawer tracks.
- Inspect the drawer rollers.

The door will not open.

- Return the ram to the top position.
- Turn off the dense pack switch.
- Push the door closed while restarting compactor.
- Check the power screws for an obstruction. See manufacturer's specifications.
- Check the power nuts for an obstruction. See manufacturer's specifications.
- Check the ram for an obstruction.

Cleaning and Deodorizing

Thoroughly clean the interior of your trash compactor regularly. It is recommend that you use a bacteria-fighting cleaner and/or degreaser to clean the ram (the platform that presses down on the garbage) and any other part of the compactor that comes into contact with the garbage.

For routine cleaning, use the following steps:

1. Always wear thick, sturdy gloves when cleaning your compactor (Figure 12-3).
2. Unplug the compactor.
3. Remove the bag and caddy, or bin, and follow the manufacturer's cleaning instructions.
4. Vacuum the inside.
5. Clean inside and outside of the compactor by using warm soapy water. Rinse and dry.
6. Close the drawer and replace the caddy with a new bag.
7. Periodically, check and replace the air freshener or charcoal filter.

Bacteria can grow on the inside of your trash compactor from the food waste that is put in the compactor. For temporary odor control between cleanings, spray the interior with a germ-killing deodorant/disinfectant. Also replace the filter (if there is one) once or twice a year.

Figure 12-3: Rubber gloves (©2006 Jupiter-Images Corporation)

General Maintenance of Hydraulic Devices

Check the compactor's preventive maintenance schedule for its most recent maintenance; also check to see if hydraulic fluid lines are adequate/intact. See the manufacturer's specifications.

Perform a Test of the Interlock Safety Device

The safety interlock prevents operation when the door is open. Test the safety device by opening the trash compactor door and press the buttons on the trash compactor to make sure it does not start up. If the compactor runs with the door open, follow the manufacturer's specifications to correct the problem (Figure 12-4).

Figure 12-4: Interlock safety device

Check the General Condition of a Dumpster and Dumpster Area

- Control litter.
- Make sure the Dumpster leasing company maintains and cleans Dumpster regularly.
- Return leaking Dumpsters for repair immediately.
- If you must wash down a Dumpster, use dry cleanup methods first, and then rinse, collect water, and discharge to appropriate drainage area.

Review Questions

1 When troubleshooting a trash compactor, what should a contractor first check?

2 When a trash compactor starts but does not complete its cycle, the contractor should check:

3 List the steps for cleaning a trash compactor.

Name: _____

Date: _____

Trash Compactor Job Sheet

1

Trash Compactor

Cleaning and Deodorizing a Trash Compactor

Upon completion of this job sheet, you should be able to clean and deodorize a trash compactor.

Task Completed

Task:

___ ① Always wear thick, sturdy gloves when cleaning your compactor.

___ ② Unplug the compactor.

___ ③ Remove the bag and caddy, or bin, and follow the manufacturer's cleaning instructions.

___ ④ Vacuum the inside.

___ ⑤ Clean inside and outside of the compactor by using warm soapy water. Rinse and dry.

___ ⑥ Close the drawer and replace the caddy with a new bag.

___ ⑦ Periodically check and replace the air freshener or charcoal filter.

Instructor's Response:

_____
_____
_____
_____

Chapter 13 Elevators

OBJECTIVES

By the end of this chapter, you will be able to:

Skill-Based

- Check and inspect floor leveling.
- Check operation of elevators.
- Perform a test on elevator doors.

Introduction

The objective of elevator maintenance is to ensure that the elevator system provides safe, dependable operation with maximum efficiency and minimum wear. A full maintenance contract from the manufacturer of the equipment installed in the building assures that they take full responsibility for that equipment. However, with a maintenance contract in place, the following should be checked and inspected to ensure the proper operation of the elevators.

Check and Inspect Elevator

The following should be reported to your supervisor:

- The platform of the elevator is not level with the floor when the door opens (Figure 13-1).
- The elevator doors cannot open fully at the destination floor for any reason (Figure 13-2).
- Unusual noises are heard during the operation of the elevator. "Look and listen" inspections will make you aware of potential troubles.

Figure 13-1: **Platform not level with floor**

Figure 13-2: **Elevator door not open fully**

With a stopwatch, check the reaction time of the elevator doors.

1. Record a baseline time. This will be the time recorded the first time you measure the time it takes for the doors to close.
2. On a regular basis, measure the time and record and compare it against the baseline time.
3. Report the results to your supervisor.

Review Questions

1. True or false? If the platform of the elevator is not level with the floor when the door opens, the technician can make the required adjustments without consulting an elevator service technician.

2. True or false? If the elevator door does not open completely for any reason, the facility maintenance technician should report the elevator as soon as possible to a qualified elevator technician or a supervisor.

Name: _____

Date: _____

Elevators Job Sheet

1

Elevators

Upon completion of this job sheet, you should be able to identify any safety requirements not being meet by the elevator.

1. Is the elevator inspected and serviced on a regular schedule? (Y/N)
2. Is the elevator capacity posted in the elevator car? (Y/N)
3. Is there an emergency phone in the elevator? (Y/N)
4. Is the turnaround space in the elevator 51 inches wide to conform to the Americans with Disabilities Act? (Y/N)

Instructor's Response:

_____
_____
_____
_____

Chapter 14 Pest Prevention

OBJECTIVES

By the end of this chapter, you will be able to:

Knowledge-Based

- Follow applicable safety procedures.

Skill-Based

- Recognize the sources of damage caused by pests.
- Select and apply proper techniques, chemicals and/or materials to eradicate and/or prevent pest infiltration.

Introduction

Pest prevention is normally not a high priority, until a problem occurs. It is not until a problem is reported that steps are taken to eliminate and then prevent the pest problems in the future. Pests are significant problems for people and property. The pesticides that are commonly used in pest control may pose potential risks to human health and the environment. You will learn how to control pests using as little pesticide as possible.

Following are steps to control and prevent pest problems.

1. Identify the pests.
2. Recognize the source.
   - Check exterior doors. If you can see light under the door, this is a potential problem.
     - Install door thresholds.
     - Seal around windows.
   - Check for windows that do not fit properly or have holes in the screens.
     - Seal around windows.
     - Use mesh or screens to fix holes in the screens.

325

- Check for openings around any objects that penetrate the building's foundation such as plumbing, electrical service, telephone wires, HVAC, etc.
  - Seal around these objects.
- Do not store materials against the foundation of a building. This could be a nesting place for bugs.
- Do not leave outside lights on all of the time. Light attracts bugs.
  - Use of motion sensor lights is an option.
- Maintain a plant-free zone of about 12 inches around the building to deter insects from entering.
- Fix problems in the structure of the building that provide a nesting place for birds or rodents.
- Place outdoor garbage containers away from the building and on concrete or asphalt slabs and keep the area clean.
- Keep facilities and area around facilities clean.

If step 2 does not eliminate the pests, continue to the next steps.

3. Identify the pests.
4. Become familiar with methods of control.
5. Estimate level of infestation.
6. Determine method of application.
7. Select pesticide for best control and least hazard.

Controlling pests will depend on the type of pests you are dealing with or the pests in your area.

Nonpesticidal Pest Control

Pest control methods such as trapping, hoeing, hand weeding, excluding the pest with barriers, sanitizing the area, and/or removing food, water, or cover for the pest.

Pest Control with Pesticides

Preventive applications of pesticides should be discouraged, and treatments should be restricted to areas of known pest activity. When pesticides are applied, the least toxic product(s) available should be used and applied in the most effective and safe manner.

Effects of Pesticides on Pests

- Stomach poison—kills when swallowed
- Contact poison—sprayed directly on pest
- Fumigants—gas inhaled or absorbed
- Systemics—will kill pest when it eats the host, but does not harm host
- Protectants—prevent pest entry

Below is a list of common pests and how to control them.

Cockroaches

Control is seldom easy because it is difficult to get the insecticide to the insect. The insecticide should have sufficient persistence to kill baby cockroaches as they hatch. If this fails, call your Environmental Health Department or pest control contractor (Figure 14-1).

Figure 14-1: Cockroach (©2006 JupiterImages Corporation)

Figure 14-2: Ants (©2006 JupiterImages Corporation)

Ants

Pour boiling water over the nest site and apply an insecticide powder. An insecticide lacquer can be applied around door thresholds or wall/floor junctions where ants run. Ant bait works so that the ant takes the bait back to the nest, killing the whole colony after a few days; Place it along where ants run (Figure 14-2).

Spiders

To remove a spider there is no need to kill it. Simply place a carton over it, then slip a piece of thin cardboard between the carton and the surface to form a lid. Take the sealed container out of the building and let the spider go (Figure 14-3).

Rodents

Seal off entry points into the facilities. Ensure areas around the facilities, including the trash container areas, are clean. Poison is available as proprietary, ready-mixed bait. Serious or persistent infestations should be dealt with by a pest control contractor or the Environmental Health Department (Figure 14-4).

Figure 14-3: Spider (©2006 JupiterImages Corporation)

Figure 14-4: Mouse (©2006 JupiterImages Corporation)

Flies (House and Fruit)

1. Inspection—locating the fly breeding and larval developmental sites.
2. Sanitation—the removal or elimination of the larval developmental sites. This step should eliminate the bulk of the fly problem so that mechanical and insecticidal measures will be more effective.
3. Mechanical controls—garbage receptacles with tight-fitting closures; tight windows and doors; windows securely screened if they can be opened; doors with self-closures; all holes through exterior walls for utilities, etc., sealed; all vents securely screened, etc.; and the use of air curtains, insect light traps, sticky-surfaced traps, etc.
4. Insecticide application—using appropriately labeled pesticides.
   - Outdoors—includes the use of boric acid in the bottom of Dumpsters.
   - Indoors—the use of automatic/metered dispensers and/or ULV applications on a room-by-room basis may be required, with the low-oil formulations being more desirable (Figure 14-5).

Figure 14-5: **Fly** (©2006 JupiterImages Corporation)

Figure 14-6: **Bees** (©2006 JupiterImages Corporation)

Stinging/Biting Insects (Bees, Mosquitoes, Wasps, Hornets, and Ticks)

Remove the pest's food, water, and shelter by keeping the facilities clean and by sanitizing outdoor areas. Keep tight-fitting lids on garbage cans and empty them regularly (Figure 14-6 through Figure 14-10).

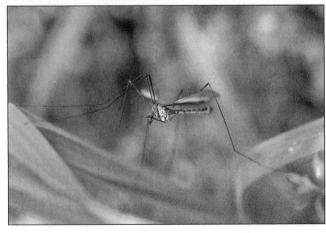

Figure 14-7: **Mosquito** (©2006 JupiterImages Corporation)

Figure 14-8: **Wasps** (©2006 JupiterImages Corporation)

Figure 14-9: Hornet (©2006 JupiterImages Corporation)

Figure 14-10: Ticks

Termites

Detecting and controlling termites is a job for the professional (Figure 14-11).

Applying Pesticides

The pesticide application equipment you use is important to the success of your pest control job. First, you must select the right kind of application equipment; then you must use it correctly and take good care of it.

Sprayers

Sprayers are the most common pesticide application equipment. They are standard equipment for nearly every pesticide applicator and are used in every type of pest control operation. Sprayers range in size and complexity from simple, hand-held models to intricate machines weighing several tons.

Figure 14-11: Termites (©2006 JupiterImages Corporation)

Hand Sprayers

Hand sprayers are often used to apply small quantities of pesticides. They can be used in structures, and they can be used outside for spot treatments or in hard-to-reach areas. Most operate on compressed air supplied by a hand pump.

- Pressurized can (aerosol sprayer)—consists of a sealed container of compressed gas and pesticides (Figure 14-12).
- Trigger pump sprayer—the pesticide and diluents are forced through the nozzle by pressure created when the trigger is squeezed. The capacity of trigger pump sprayers ranges from 1 pint to 1 gallon (Figure 14-13).
- Hose-end sprayer—causes a fixed rate of pesticide to mix with the water flowing through the hose to which it is attached (Figure 14-14).
- Push-pull hand pump sprayer—works with a hand-operated plunger that forces air out of a cylinder, creating a vacuum at the top of a siphon tube. The suction

Figure 14-12: Pesticide aerosol can

Figure 14-13: Trigger pump pesticide sprayer

Figure 14-14: **Spraying with hose-end pesticide sprayer** (©2006 JupiterImages Corporation)

Figure 14-15: **Push-pull hand sprayer pump**

draws pesticide from a small tank and forces it out with the air flow. Capacity is usually 1 quart or less (Figure 14-15).
- Compressed air sprayer—usually a hand-carried sprayer that operates under pressure created by a self-contained manual pump (Figure 14-16).

Small Motorized Sprayers

Small motorized sprayers are usually not self-propelled. They may be mounted on wheels so they can be pulled manually, mounted on a small trailer for pulling behind a small tractor, or skid-mounted for carrying on a small truck. They may be low-pressure or high-pressure, according to the pump and other components with which they are equipped.

- Estate sprayers—These sprayers are mounted on a two-wheel cart with handles for pushing (Figure 14-17).
- Power backpack sprayer—This backpack-type sprayer has a small gasoline-powered engine. This model can generate high pressure and is best suited for low-volume applications of diluted or concentrated pesticide (Figure 14-18).

Figure 14-16: Compressed air pesticide sprayer

Figure 14-17: Estate pesticide sprayer

Follow Applicable Safety Procedures

When working with pesticides, use the following procedures:

- Choose the lowest toxicity pesticide that can be used legally on the target area, crop, or plant and that will safely and effectively control the pest.
- Plan ahead and buy no more pesticide than you need.
- Keep pesticides separate from other items.
- Make sure you have the proper safety and application equipment available and know how to use it.
- Read, understand, and follow the pesticide's label directions.
- Examine the area to be treated and the surrounding area. If there are plants or animals that could be harmed by the pesticide, don't spray if you cannot guarantee they will not be injured.
- Never put potentially hazardous waste, such as pesticides, directly in the garbage or pour remaining chemicals down the drain. Check if your community has a household hazardous waste collection program.
- Store pesticides out of reach of children—in locked cabinets or in cabinets with childproof latches.
- Store pesticides only in their original containers with labels visible and intact.

Figure 14-18: Power backpack sprayer

Pests of any kind are a nuisance to tenants. When working on a complaint about pests, be as nonintrusive as possible, but do whatever is necessary to eliminate the pests. Once the elimination or preventive measures have been performed, check back with the tenant to make sure the pests have been eliminated.

Review Questions

1. Describe the difference between pest control with pesticides and nonpesticidal pest control.

2. Describe how to control the following pests:

   Cockroaches

   Ants

   Spiders

   Termites

Name: _____

Date: _____

Pest Prevention Job Sheet 1

Pest Prevention

Identify and Remove Pests

Upon completion of this job sheet, you should be able to identify pests that may be causing problems at your facility and develop a plan to remove them.

1. Inspect your facility or training area and make a list of the pests that you observe or pests that the tenants report to you. If possible, use a digital camera to photograph the pests for easier identification.

2. Document where the pests were observed and what they were doing at the time of observation.

3. Estimate the level of infestation to determine if there needs to be a plan in place to remove the pests.

4. Research the appropriate method for removing the pests if deemed necessary and develop a plan.

5. Implement your plan if applicable.

Instructor's Response:

_____
_____
_____
_____

Chapter 15 Groundskeeping

OBJECTIVES

By the end of this chapter, you will be able to:

Skill-Based

- Maintain and police grounds including mowing, edging, planting, mulching, leaf removal, and other assigned tasks.
- Perform basic small engine repair and preventive maintenance according to manufacturer's specifications.
- Perform basic swimming pool maintenance not requiring certification.
- Remove refuse and snow as required.
- Maintain public areas including hallways, kitchens, and lobbies.
- Repair asphalt by using cold-patch material.

Introduction

Groundskeeping is the activity of tending an area for aesthetic or functional purposes. It includes mowing grass, trimming hedges, pulling weeds, planting flowers, and so on.

Landscaping and groundskeeping workers maintain grounds by using hand or power tools or equipment. Workers typically perform a variety of tasks, which may include: sod laying, mowing, trimming, planting, watering, fertilizing, digging, raking, sprinkler installation, and installation of mortarless segmental concrete masonry wall units.

Mowing

Depending on the size of the lawn, most technicians use either a push mower (which may be self-propelled) or a riding mower (Figure 15-1).

Riding mowers generally come in several different types. The most common riding mower for residential use is one with a "belly" deck. A **belly deck** is a mower that is built like a car (four wheels, a driver's seat, a steering

Figure 15-1: Push mower (©2006 Jupiter-Images Corporation)

Figure 15-2: Riding mower

wheel, engine located in the front) and has a blade deck underneath the driver and engine (Figure 15-2).

During growing season, the following recommended procedures should be followed:

- Mow the grass once a week.
- Mower wheel setting should be 2½–3 inches high.
- Growing season is usually May to October.
- Mowing does not have to be done outside of growing season, unless to help with leaf clean-up in late fall.

Edging

Edging or trimming is the final step in mowing your lawn. It's a simple but all-important practice that gives a lawn a finished, manicured look. Edging can be done using the following:

- Use a handheld edger between pavement and grass. Place the wheel on the pavement with the blade over the edge and push and pull. For large lawns, use a power-driven model (Figure 15-3).
- Use grass shears around trees, the edges of beds, or in places that are hard to reach (Figure 15-4).
- Use a string trimmer to trim and edge large lawns or to cut grass too tall to mow (Figure 15-5).

Edging information:

- Edge once every other week along with trimming around landscape with a gas trimmer.

Figure 15-3: Handheld edger

CHAPTER 15 Groundskeeping

Figure 15-4: Grass shears (©2006 JupiterImages Corporation)

Figure 15-5: String trimmer

- Edging is done along walkways to prevent grass from growing out past the borders.
- Edging is done along edges of turf and landscape beds to prevent grass from growing into the beds.

Mulch

Mulching trees and shrubs is a good method to reduce landscape maintenance and keep plants healthy. Mulch helps conserve moisture, providing 10 to 25 percent reduction in soil moisture loss from evaporation. Mulch helps keep the soil well aerated by reducing soil compaction that results when raindrops hit the soil. It also reduces water runoff and soil erosion.

Mulch is any material applied to the soil surface for protection or improvement of the area covered. Nature produces large quantities of mulch all the time with fallen leaves, needles, twigs, pieces of bark, spent flower blossoms, fallen fruit, and other organic material. There are basically two types of mulches: organic and inorganic. Both types may have their place in the landscaping.

An **organic mulch** is a mulch made of natural substances such as bark, wood chips, leaves, pine needles, or grass clippings. Organic mulches attract insects, slugs, cutworms, and the birds that eat them. They decompose over time and need to be replaced after several years. **Inorganic mulches**, such as gravel, pebbles, black plastic, and landscape fabrics, do not attract pests and they do not decompose. A 2- to 4-inch layer (after settling) is adequate to prevent most weed seeds from germinating. Mulch should be applied to a weed-free soil surface (Figure 15-6).

Figure 15-6: Mulch around plants (©2006 JupiterImages Corporation)

Using Bushes and Shrubs in Landscapes

Flowering shrubs and evergreen bushes both have their place in landscaping. Many flowering shrubs attract birds with their berries and provide both fall foliage and winter interest. Evergreen bushes can be pruned into hedges or can function as privacy screens. Either flowering shrubs or evergreen bushes can stand alone as specimens used for focal points (see Figures 15-7 and 15-8).

Planting Bushes and Shrubs

1. Dig receiving hole as per instruction on bushes or shrubs.
2. Place shrub in hole and back fill while watering in the shrub.
3. Set irrigation to water 2 gallons per week.
4. Mulching the beds is done mid-spring, usually with bark or woodchips.
5. Mulch 2–3 inches deep helps keep weeds to a minimum.

Figure 15-7: Flowering shrubs (©2006 JupiterImages Corporation)

Winterizing the Irrigation System

Every year, before the first freeze, irrigation "blow-out" becomes the priority for all irrigation systems that are in parts of the country where the frost level extends below the depth of the installed piping. To minimize the risk of freeze damage to your irrigation system, you'll need to "winterize" your irrigation system.

Manual Drain Method

This method is used when the manual valves are located at the end and low points of the irrigation piping (Figure 15-9).

To drain these systems:

1. Shutoff the irrigation water supply. The shutoff will be located in the basement and will be either a gate/globe valve, ball valve, or stop and waste valve (Figure 15-10 through Figure 15-12).
2. Open all the manual drain valves.
3. Once the water has drained out of the mainline, open the boiler drain valve or the drain cap on the stop and waste valve and drain all the remaining water that is between the irrigation water shutoff valve and the backflow device (Figure 15-13).
4. Open the test cocks on the backflow device. If your sprinklers have check valves, you will need to pull up on the sprinklers to allow the water to drain out the bottom of the sprinkler body (Figure 15-14).

Figure 15-8: Evergreen bushes (©2006 JupiterImages Corporation)

CHAPTER 15 *Groundskeeping* **339**

Figure 15-9: Valves on the irrigation piping

Figure 15-10: Gate valve
(©2006 JupiterImages Corporation)

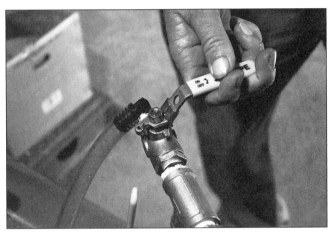

Figure 15-11: Ball valve

Figure 15-12: Stop/waste valve

Figure 15-13: Boiler drain valve/drain cap

Figure 15-14: Test cocks

Automatic Drain Method

This method is used when automatic drain valves are located at the end and low points of the irrigation piping (Figure 15-15). These will automatically open and drain water if the pressure in the piping is less than 10 psi.

1. To activate the automatic drain valves, shut off the irrigation water supply and activate a station to relieve the system pressure. The shutoff will be located in the basement and will be either a gate/globe valve, ball valve or stop and waste valve (Figure 15-16 through Figure 15-18).
2. Once the water has drained out of the mainline, open the boiler drain valve or the drain cap on the stop and waste valve and drain the remaining water that is between the irrigation water shutoff valve and the backflow device (Figure 15-19).
3. Open the test cocks on the backflow device. If your sprinklers have check valves, you will need to pull up on the sprinklers to allow the water to drain out the bottom of the sprinkler body (Figure 15-20).

Figure 15-15: Valves at end of irrigation piping

Figure 15-16: Gate valve (©2006 JupiterImages Corporation)

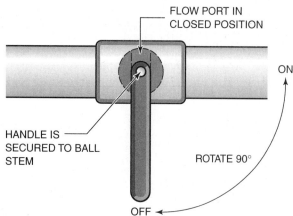

Figure 15-17: Ball valve

Figure 15-18: Stop/waste valve

CHAPTER 15 *Groundskeeping* 341

Figure 15-19: **Boiler drain valve/drain cap**

Figure 15-20: **Test cocks**

"Blow-Out" Method

It is recommended that a qualified licensed contractor perform this type of winterization method. The blow-out method utilizes an air compressor with a cubic foot per minute (CFM) rating of 125–185 for any mainline of 2 inches or less and a PSI of 50–80.

1. Open the test cocks on the vacuum breaker (Figure 15-21).
2. Shutoff the irrigation water supply and open the drain on the supply line.
3. Once the line is drained, close the drain.
4. Attach the compressor to the mainline via a quick coupler, hose bib, or other type of connection, which is located before the backflow device (Figure 15-22).
5. Activate the station on the controller that is the zone or sprinklers highest in elevation and the furthest from the compressor (Figure 15-23).

Figure 15-21: **Test cocks**

Figure 15-22: **Compressor attached to the mainline**

Figure 15-23: **Station controller**

6. Do not close the backflow isolation or test cock valves. Slowly open the valve on the compressor; this should gradually introduce air into the irrigation system. The air pressure should be constant at 50 psi. If the sprinkler heads do not pop up and seal, increase the air until the heads do pop up and seal. The air pressure should NEVER exceed 80 psi.
7. Activate each station/zone starting from the furthest station/zone from the compressor, slowly working your way to the closest station/zone to the compressor. Each station/zone should be activated until no water can be seen exiting the heads. This should take approximately 2 to 4 minutes per station/zone.

Spring Irrigation Startup

If the system was correctly winterized in the fall, the chances of cracks and breaks due to freezing were greatly reduced. But even properly winterized systems are subject to damage from extreme conditions.

Spring startup procedures basically consist of four phases:

- safely reintroducing water to each zone
- checking winter damage, making repairs as needed, and cleaning or replacing nozzles
- examining the entire system to see that it is still operating the way it is supposed to and providing even coverage
- resetting the controls

Starting Up the Irrigation System for Spring

1. Before turning on any water to the system, make sure all manual drain valves are returned to the CLOSED position (Figure 15-24).
2. Open the system main water valve slowly to allow pipes to fill with water gradually. If these valves are opened too quickly, sprinkler mainlines are subjected to high surge pressures, uncontrolled flow, and water hammer.
3. Verify the proper operation of each zone valve by manually activating it from the controller.

Figure 15-24: Manual drain valve

4. Activate each station on the controller, checking for proper operation of the zone. Check for proper operating pressure (low pressure indicates a line break or missing sprinkler), proper rotation and adjustment of sprinkler heads, and adequate coverage. Check and clean filters on poorly performing sprinklers. Adjust heads to grade as necessary.
5. Reprogram the controller for automatic watering. Replace the controller backup battery if necessary.
6. Uncover and clean the system rain sensor, if applicable (Figure 15-25).
7. Finish and clean any in-line filters for drip irrigation zones.

Aeration

Lawn aeration involves the removal of small soil plugs or cores out of the lawn. Although hand aerators are available, most aeration is done mechanically with a machine having hollow tines or spoons mounted on a disk or drum. Known as a core aerator, it extracts ½- to ¾-inch diameter cores of soil and deposits them on your lawn. Aeration holes are typically 1–6 inches deep and 2–6 inches apart. Other types of aerators push solid spikes or tines into the soil without removing a plug (spiking). These are not as effective because they can contribute to compaction. Core aeration is a recommended lawn care practice on compacted, heavily used turf and to control thatch build-up (Figure 15-26).

Figure 15-25: **Rain sensor**

Aeration should be done twice a year, spring and fall. Mark sprinkler heads with flag or paint before aeration.

1. The soil should be moist but not wet.
2. Lawns should be thoroughly watered two days prior to aerating, so tines can penetrate deeper into the soil and soil cores easily fall out of the tines. If aerating after prolonged rainfall, it is important to wait until the soil has dried somewhat so soil cores do not stick in the hollow tines.
3. Aerate the lawn in at least two different directions to ensure good coverage. Be careful on slopes, especially steep ones, as well as near buildings and landscape beds.

Figure 15-26: **Aeration machine**

Pool Maintenance

Routine pool maintenance is an unavoidable fact of life if you want to keep your pool looking clean, sparkling, and inviting day after day.

Cleaning the Pool Deck

1. Remove as much debris as possible from the pool.
2. Sweep or use a hose to remove the debris near the pool. Cover pool if necessary (Figure 15-27).

Figure 15-27: Pool deck (©2006 JupiterImages Corporation)

Cleaning the Surface of the Pool

Dirt floating on the surface of the water is easier to remove than it is from the bottom. Remove floating debris off the surface by using a leaf rake and telepole. As the net fills, empty it into a trash can or plastic garbage bag. Do not empty your skimming debris into the garden or on the lawn for the debris is likely to blow right back into the pool as soon as it dries out.

There is no particular method to skim, but as you do, scrape the tile line, which acts as a magnet for small bits of leaves and dirt. The rubber-plastic edge gasket on the professional leaf rake will prevent scratching the tile (Figure 15-28).

If there is scum or general dirt on the water surface, squirt a quick shot of tile soap over the length of the pool. The soap will spread the scum toward the edges of the pool, making it more concentrated and easier to skim off.

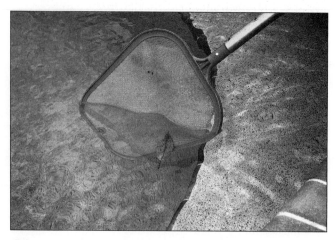

Figure 15-28: Skim debris from the pool

Maintaining the Pool

1. Sanitize your pool with a stabilized chlorine product to provide protection against bacteria. These generally come in stick or tablet form and are fed into a distribution container near the pump and filter system.
2. Use an algae preventive or inhibitor to help keep the more than 15,000 kinds of algae from ever getting started. This liquid product is simply poured into the water near the skimmer intake so that the pump system can distribute it to all areas of the pool.
3. Find a pool professional or supply dealership that has a water test facility, or access to one, and that offers computer analysis of samples you bring in.
4. Check for chlorine and PH levels daily. To check PH in pool water, use a kit from your local pool store. It contains tubes that hold about 4 ounces of pool water. You then put in drops of liquid that comes with the test kit, shake for about 30 seconds, and then match the colors with the accompanying chart to check the chlorine and pH levels (yellow for chlorine and red for pH).

Snow Plowing

Unfortunately, as with rainfall, you cannot always predict when or how much snow will fall in any one spot. When it does fall, it is the responsibility of the facilities maintenance technician to remove the snow. This does not mean that the technician has to remove the snow herself; the snow removal service can be contracted out.

Recommended Snow Removal Procedures

- Steps, large walk-throughs and large entry ways shall be partially shoveled with a path along the railings for initial opening of these areas. Handicap areas must be fully accessible (Figure 15-29).
- Clean-up operations after a storm shall involve completion of opening all walks and entry ways and de-icing.
- Snow shall be pushed back from sides of roadways, walks, and parking lots (Figure 15-30).

Figure 15-29: Clear walkways

Figure 15-30: Snow on side of parking lot

- Stairs and entry ways shall have all remaining snow removed.
- Ice choppers and ice-melt applications will then be employed on remaining ice.
- Application of ice-control products will follow plowing, based on current weather conditions, or when freezing occurs, as determined by the grounds supervisor.
- Return trips to remove melted ice and slush to the surface will complete the cleaning of all surfaces.
- Return trips for sanding and salting equipment shall be made as often as necessary on roads, as determined by the grounds supervisor.

When parking lots are plowed, snow should be piled so as not to block thoroughfares and sidewalk areas. If snow has to be pushed up over a curbed area, it should be piled so that it will not fall back into the lot and will still be clear of any adjacent sidewalks.

Small Engine Repair

The facilities maintenance technician should be able to perform basic small engine repair and preventive maintenance according to manufacturer's specifications.

Common Problems with Small Engines

Engine Will Not Start

Figure 15-31: Spark plug (©2006 JupiterImages Corporation)

- Check for proper fuel level—Fill gas tank to just below the fill neck so there is room for the gas to expand and slosh around. If you fill the tank all the way to the top, then gas will leak out as the engine shakes around.
- Check to see if the fuel valve is in the ON position.
- Check the spark plug—If it is "carbon shorted," carbon between the electrode gap, clean or replace it. If the plug is pitted, burned, or has cracked porcelain, replace it with an identical replacement spark plug (Figure 15-31).
- Check for spark—With a commercially available spark tester, test for spark by putting the spark plug wire (high-tension lead) on one side of the tester and clip the other side of the tester to the shroud, fins, or head bolts (anything metallic). Or ground the plug, with high tension lead on it, to the head of the engine and crank on the engine. Spark is present when you see blue sparks jump the electrode gap.
- Prime or choke the engine—If the engine has a primer or choke, use them! Prime the engine three times and then start. Choke engine until it starts to fire and then open the choke.

If after going through these steps the engine still won't start, take it in to a repair center to be repaired.

Service Recommendations

- Change the oil—This is the most important thing you can do to make your engine live forever. Change your oil with a good grade of SAE 30w HD for Service SC and higher. Do not overfill—this will cause the engine to smoke or blow oil out of the breather into the air filter. On an engine with a dipstick, fill to the top line of the crosshatch mark. On engines without dipsticks, fill to the top of the hole. On horizontal engines without dipsticks, fill until oil comes out of the fill hole.
- Clean or replace the air filter—This is also a very important part of engine maintenance. Clean the foam filters with soap and hot water. Re-oil and squeeze out excess. Paper elements must be replaced if extremely dirty of have dirt caked up in the pleats (Figure 15-32).
- Change spark plugs—Replace them once a season instead of trying to clean them and reuse them (Figure 15-33).
- Remove fuel before storing engine—Doing this will prevent unwanted maintenance in the spring time.

If, after you have completed all of these steps, the engine still will not start, take it in to a repair center to be repaired.

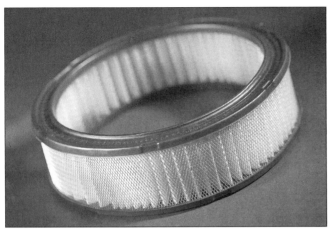

Figure 15-32: Air filter (©2006 JupiterImages Corporation)

Figure 15-33: Spark plug (©2006 JupiterImages Corporation)

Engine Smokes

- Dirty or plugged air filter—Replace or clean air filter (Figure 15-34).
- Wrong grade of oil—Oil that is less than 30w HD will vaporize if used for a long time. Use only use a good grade of 30w HD oil for all four-cycle lawn equipment.
- Worn valve guides—A valve guide that is worn out due to excessive wear on the engine and must be taken to the shop for a replacement guide bushing. Common causes are mowing on a hillside for excessive amounts of time and debris build-up on the cooling fins (Figure 15-35).
- Choke is still on—Open the choke as soon as the engine fires and continues to run without faltering.
- Too much oil—Correct the amount of oil in the crankcase. Fill only to the FULL mark.

If after going through these steps the engine still will not start, take it in to a repair center to be repaired.

Engine Sputters

- Some water is in the gas—Remove this old fuel and replace with fresh mid-grade gasoline.

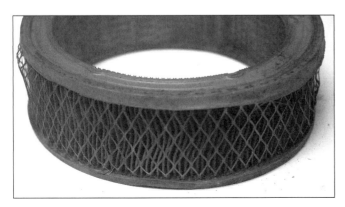

Figure 15-34: Dirty air filter

Figure 15-35: Valve guide

Figure 15-36: Flywheel key

Figure 15-37: Flywheel

Figure 15-38: Bad spark plug

Figure 15-39: Good spark plug (©2006 JupiterImages Corporation)

- Electrical system is grounding out on the equipment (not engine). Check for cracks for exposed wires. Repair as necessary.
- Flywheel key sheared—Replace key (Figure 15-36).
- Stop wire is rubbing on flywheel—Remove flywheel, tape up wire if not severed, and re-route it so it will not touch flywheel (Figure 15-37).
- Bad spark plug (Figure 15-38)—Replace the spark plug with an identical new one (Figure 15-39).

If after going through these steps the engine still will not start, take it in to a repair center to be repaired.

Maintain Public Areas

Public areas are defined as areas not assigned to individuals. This includes kitchens, hallways, lobby areas, and stairwells—areas open to the public.

Recommended Maintenance Procedures for Public Areas

Bathroom

- disinfected on a daily basis
- paper products restocked on a daily basis
- thoroughly cleaned once a week
- hand soap replaced as needed

Stairways

- thoroughly cleaned at least once a week
- swept on a daily basis

Hallway

- buffed to a shine at least once a week, twice time permitting
- general cleaning on a daily basis
- walls cleaned weekly and spot cleaned as needed

- alcoves and shelves thoroughly cleaned
- water fountains disinfected daily

Trash Cans
- emptied on a daily basis
- disinfected and thoroughly cleaned once a week

Lobby
- thoroughly cleaned weekly
- general cleaning on a daily basis

Porches, Patios, and Dumpster Pads
- broom swept as needed

Repair Asphalt by Using Cold-Patch Material

Although asphalt driveways and parking lots are fairly durable, they do require some maintenance, especially in cold areas where freeze/thaw cycles are the norm. Minor damage caused by water getting into small cracks and then freezing can quickly become major problems. Periodic sealing will help keep these cracks from starting, but they will not prevent damage caused by settling of the ground under the driveway or improper installation.

Figure 15-40: Use chisel or sharp object to clean out the crack (©2006 JupiterImages Corporation)

Cold patch is a fast, permanent, easy-to-use repair material for asphalt and concrete surfaces.

Repairing Small Cracks

1. Fill any cracks in a blacktop drive as soon as possible to keep water from getting under the slab and causing more serious problems. Cracks that are ½ inch and wider are filled with asphalt cold patch, sold in bags and cans. Narrow cracks are treated with crack-filler, which is available in cans, plastic pour bottles, and handy caulking cartridges.
2. Use a masonry chisel, wire brush, or similar tool to dig away chunks of loose and broken material from the crack (Figure 15-40).
3. Sweep out the crack with a stiff-bristled broom (Figure 15-41).
4. Use a garden hose with a pressure nozzle to clean off all dust.

Figure 15-41: Sweep out the crack with stiff-bristled broom (©2006 JupiterImages Corporation)

5. Apply the crack-filler. For repairs deeper than 2 inches, apply and tamp cold patch in 1- to 2-inch layers. Add enough material so that tamping or tire rolling leaves a slight crown. If over time a patch settles below the road surface, clean the surface of the patch, add another layer, and compact (Figure 15-42).

Figure 15-42: Apply crack-filler to the crack

Repair Large Holes with Cold Patch

Large holes or potholes require asphalt cold patch, which has larger aggregate than paste patches. For this application, purchase 60- to 70-pound bags. Cold patch resembles driveway asphalt but is treated with chemicals to keep it workable. If the temperature stays above 50°F, you don't have to heat it the way highway crews heat fresh asphalt—but you do have to roll it.

1. Clear loose debris, and remove jagged edges around the hole with a hammer and cold chisel (Figures 15-43 and 15-44).

Figure 15-43: Clear loose debris

Figure 15-44: Use hammer or cold chisel to remove jagged edges around a hole

2. Pile on enough cold patch to leave a slight mound after tamping (Figure 15-45).
3. Fill deep holes in two stages, tamping in between, to avoid leaving a water-collecting depression.
4. Apply the weight of your car, driving it slowly over a piece of ¾-inch plywood or a layer of sand spread over the cold patch (Figure 15-46).

Figure 15-45: **Cold patch piled in a hole**

Figure 15-46: **Drive a car over the cold patch to spread it**

Review Questions

1. List the steps for planting brushes and shrubs.

2. List three methods for winterizing an irrigation system.

3. List the steps for starting up the irrigation system in the spring.

4. List the steps for maintaining a swimming pool.

5. If the lawnmower's engine does not start, what should the technician check?

6. What should the technician check if the engine on the lawnmower starts to sputter?

Name: _____

Date: _____

Groundskeeping Job Sheet

1

Groundskeeping

Lawnmower Maintenance

Upon completion of this job sheet, you should be able to identify a need for mower maintenance.

Task Completed

___ ❶ Use the gas and oil recommended by the manufacturer.
___ ❷ Blades are sharp.
___ ❸ Blades and crankshaft are tight.
___ ❹ Underside of mower is cleaned after each use.
___ ❺ Check grass-catcher bag for wear and tear or deterioration.
___ ❻ Check mower wheels, bearings, and axles for wear and lubrication.
___ ❼ Mower is thoroughly inspected every year.

Instructor's Response:

_____
_____
_____
_____

Appendix

Conversion Tables

Fraction	Decimal	Fraction	Decimal	Fraction	Decimal
1/64	0.0156	11/32	0.3438	23/32	0.7188
1/32	0.0313	23/64	0.3594	47/64	0.7344
3/64	0.0469	**3/8**	**0.3750**	**3/4**	**0.7500**
1/16	0.0625	25/64	0.3906	49/64	0.7656
5/64	0.0781	13/32	0.4063	25/32	0.7813
3/32	0.0938	27/64	0.4219	51/64	0.7969
7/64	0.1094	7/16	0.4375	13/16	0.8125
1/8	**0.1250**	29/64	0.4531	53/64	0.8281
9/64	0.1406	15/32	0.4688	27/32	0.8438
5/32	0.1563	31/64	0.4844	55/64	0.8594
11/64	0.1719	**1/2**	**0.5000**	**7/8**	**0.8750**
3/16	0.1875	33/64	0.5156	57/64	0.8906
13/64	0.2031	17/32	0.5313	29/32	0.9063
7/32	0.2188	35/64	0.5469	59/64	0.9219
15/64	0.2344	9/16	0.5625	15/16	0.9375
1/4	**0.2500**	37/64	0.5781	61/64	0.9531
17/64	0.2656	19/32	0.5938	31/32	0.9688
9/32	0.2813	39/64	0.6094	63/64	0.9844
19/64	0.2969	**5/8**	**0.6250**		

Conversion Factors

There are two basic systems used to make measurements in the world today. They are the **English system** and **metric system.** The English system of measure is the unit of measure currently used in the United States today by most technicians. Its base units are inch (in), foot (ft), yard (yd), and mile (mi).

Unit	Divisions
	English
1 inch (in)	
1 foot (ft)	12 in
1 yard (yd)	3 ft
1 mile (mi)	5,280 ft

English conversion factors

The base unit for the metric system is the meter. This base unit is further divided into larger and smaller units (in multiples of 10) by adding prefixes. Common metric prefixes are deci- (10), centi- (100), and milli- (1000). For example, a meter contains 10 decimeters.

Unit	Divisions
Units of Length	
10 millimeters (mm)	= 1 centimeter (cm)
10 centimeters	= 1 decimeter (dm) = 100 millimeters
10 decimeters	= 1 meter (m) = 1000 millimeters
10 meters	= 1 dekameter (dam)
10 dekameters	= 1 hectometer (hm) = 100 meters
10 hectometers	= 1 kilometer (km) = 1000 meters

Metric conversion factors

The metric system of measure is occasionally used in the United States and therefore it is important that the technician be able to recognize and convert from one unit to another. The table provided below gives conversion factors for converting from one system to another.

Starting with	Multiply	To Find
inches	2.5	centimeters
feet	30	centimeters
yards	0.9	meters
miles	1.6	kilometers
centimeters	0.4	inches
centimeters	0.0333	inches
meters	1.1111	yards
kilometers	0.625	miles

Converting from Metric to English and English to Metric

For example, to convert 12 feet 6 inches to centimeters, the technician would perform the following steps.

Convert from inches to centimeters

Step 1 Converting 12 feet into inches.

$$12\,ft \times 12\,in/ft = \mathbf{144\,in}$$

Step 2 Next add the inch portion of the original measurement.

$$144\,in + 6\,in = \mathbf{120\,in}$$

Step 3 Find the conversion factor using the table above.

$$120\,in \times 2.5\,cm/in = \mathbf{300\,cm}$$

Convert from centimeters to inches

Step 1 Find the conversion factor using the table above.

$$300\,cm \times 0.4\,in/cm = \mathbf{120\,in}$$

Step 2 Convert from inches to feet and inches by dividing by 12.

$$120\,in \,/\, 12\,ft/in = 12\,ft \text{ with a remainder of } 6$$

Therefore, the measure would be written as **12 ft 6 in or 12′6″**.

Electrical Wire Gauge and Current Chart

AWG Gauge	Diameter Inches	Ohms/1000 ft	Maximum Amps
0000	0.46	0.049	302
000	0.4096	0.0618	239
00	0.3648	0.0779	190
0	0.3249	0.0983	150
1	0.2893	0.1239	119
2	0.2576	0.1563	94
3	0.2294	0.197	75
4	0.2043	0.2485	60
5	0.1819	0.3133	47
6	0.162	0.3951	37
7	0.1443	0.4982	30
8	0.1285	0.6282	24
9	0.1144	0.7921	19
10	0.1019	0.9989	15
11	0.0907	1.26	12
12	0.0808	1.588	9.3
13	0.072	2.003	7.4
14	0.0641	2.525	5.9
15	0.0571	3.184	4.7
16	0.0508	4.016	3.7
17	0.0453	5.064	2.9
18	0.0403	6.385	2.3

Small Engine Recommended Preventive Maintenance Charts

Recommended Chainsaw Preventive Maintenance				
	Daily	Weekly	Monthly	As Needed
Sprocket	Inspect			Replace
Fuel filter		Clean		
Muffler			Clean	
Muffler screen	Clean			Replace
Fuel tank			Clean	
Spark plug		Clean and adjust		Replace
Fuel, oil, and hoses	Check			
Air filter	Clean	Replace		
Screws, nuts, bolts	Inspect and tighten			
Chain	Inspect and sharpen			

Recommended Lawnmower Preventive Maintenance				
	Daily	Weekly	Monthly	As Needed
Fuel filter		Clean		
Muffler			Clean	
Fuel tank			Clean	
Spark plug		Clean and adjust		Replace
Fuel, oil, and hoses	Check			
Air filter	Clean	Replace		
Screws, nuts, bolts	Inspect and tighten			

Recommended String Trimmer Preventive Maintenance				
	Daily	Weekly	Monthly	As Needed
Fuel filter		Clean		
Muffler			Clean	
Fuel tank			Clean	
Spark plug		Clean and adjust		Replace
Fuel, oil, and hoses	Check			
Air filter	Clean	Replace		
Screws, nuts, bolts	Inspect and tighten			

Glossary

Accessibility Ability to go in, out, and through a building and its rooms with ease regardless of disability.

Accessible route The walking surface from the exterior access through the residence that is required to connect all spaces that are part of the dwelling unit. If only one route is provided, it cannot pass through a bathroom, closet, or similar space.

Acoustics The science of sound and sound control.

Active solar system System that uses mechanical devices to absorb, store, and use solar heat.

Adobe Heavy clay used in many southwestern states to make sun-dried bricks.

Aggregate Stone, gravel, cinder, or slag used as one of the components of concrete.

Air duct A pipe, typically made of sheet metal, that carries air from a source such as a furnace or air conditioner to a room within a structure.

Air trap A U-shaped pipe placed in wastewater lines to prevent backflow of sewer gas.

Air-dried lumber Lumber that has been stored in yards or sheds for a period of time after cutting. Building codes typically assume a 19 percent moisture content when determining joists and beams of air-dried lumber.

Alcove A small room adjoining a larger room often separated by an archway.

Alternative braced wall panel (ABWP) A method of bracing a braced wall line that uses panels with a minimum length of 2'-8" (800 mm) to resist lateral loads.

Ampere (amp) A measure of electrical current.

Anchor A metal tie or strap used to tie building members to each other.

Anchor bolt A threaded bolt used to fasten wooden structural members to masonry.

Angle iron A structural piece of steel shaped to form a 90° angle.

Apron The inside trim board placed below a windowsill. The term is also used to refer to a curb around a driveway or parking area.

Areaway A subsurface enclosure to admit light and air to a basement. Sometimes called a window well.

Asbestos A mineral that does not burn or conduct heat; it is usually used for roofing material.

Ash dump An opening in the hearth where ashes can be dumped.

Ash pit An area in the bottom of the firebox of a fireplace to collect ash.

Ashlar masonry Squared masonry units laid with a horizontal bed joint.

Asphalt An insoluble material used for making floor tile and for waterproofing walls and roofs.

Asphalt shingle Roof shingles made of asphalt-saturated felt and covered with mineral granules.

Asphaltic concrete A mixture of asphalt and aggregate that is used for driveways.

Assessed value The value assigned by governmental agencies to determine the taxes to be assessed on structures and land.

Atrium An inside courtyard of a structure that may be either open at the top or covered with a roof.

Attic The area formed between the ceiling joists and rafters.

Awning window A window that is hinged along the top edge.

Backfill Earth, gravel, or sand placed in the trench around the footing and stem wall after the foundation has cured.

Baffle A shield, usually made of scrap material, used to keep insulation from plugging eave vents. Also used to describe wind- or sound-deadening devices.

Balance A principle of design dealing with the relationship between the various areas of a structure as they relate to an imaginary centerline.

Balcony An above-ground deck that projects from a wall or building with no additional supports.

Balloon framing (Eastern framing) A construction method that has vertical wall members that extend uninterrupted from the foundation to the roof.

Balusters One of a series of closely spaced, ornamental, vertical supports for a railing.

Balustrade A low ornamental railing used on the roofs of Georgian and Federal style homes to surround a flattened central area of a low-pitched hipped roof, forming what is often referred to as a "widow's walk" (from which sea captains' wives were supposed to have watched for their husbands' ships). Balustrades on Federal houses are usually found above the exterior walls.

Band joist (Rim joist) A joist set at the edge of the structure that runs parallel to the other joist.

Banister A handrail beside a stairway.

Barge rafter The inclined trim that hangs from the projecting edge of a roof rake.

Base cabinets Cabinets in a kitchen or bathroom that sit on the floor.

Base course The lowest course in brick or concrete masonry unit construction.

Baseboard The finish trim where the wall and floor intersect, or an electric heater that extends along the floor.

Baseline A reference line in mapping.

Basement A level of a structure that is built either entirely below grade level (full basement) or partially below grade level (daylight basement).

Basement wall The portion of a wall that is partially or totally below grade and encloses a basement.

Batt Insulation usually made of fiberglass to be used between framing members.

Battlement A parapet wall with open spaces for shooting.

Batten A board used to hide the seams when other boards are joined together.

Bay window A window placed in a projection of an exterior wall that extends all the way down to the foundation. In plan view, the wall projection may be rectangular, polygonal, or curved. See also Oriel window.

Beam A horizontal structure member that is used to support roofs or wall loads (often called a header).

Beamed ceiling A ceiling that has support beams that are exposed to view.

Bearing plate A support member, often a steel plate, used to spread weight over a larger area.

Bearing wall A wall that supports vertical loads in addition to its own weight.

Beveled siding Siding that has a tapered thickness.

Bibb An outdoor faucet that is threaded so that a hose may be attached. (Represented by *H.B.* on floor plans.)

Bird block (Eave blocking) A block that is placed between rafters or trusses to maintain a uniform spacing and to keep animals out of the attic.

Bird's mouth A notch cut into a rafter to provide a bearing surface where the rafter intersects the top plate.

Blind nailing Driving nails in such a way that the heads are concealed from view.

Blocking (Bridging) Framing members, typically wood, that are placed between joists, rafters, or studs to provide rigidity.

Board and batten A type of siding using vertical boards with small wood strips (battens) used to cover the joints of the boards.

Board foot The amount of wood contained in a piece of lumber 1 inch thick by 12 inches wide by 12 inches long (25 × 300 × 300 mm).

Bolt, anchor An L-shaped bolt used to connect wood members to concrete.

Bolt, carriage A bolt with a rounded head and a threaded shaft used for connecting steel and other metal members as well as timber connections.

Bolt, machine A bolt with a hexagonal head and a threaded shaft used to attach steel to steel, steel to wood, or wood to wood.

Bond The mortar joint between two masonry units, or a pattern in which masonry units are arranged.

Bond beam A reinforced concrete beam used to strengthen masonry walls.

Bottom chord The lower, usually horizontal, member of a truss used to support the ceiling material.

Box beam A hollow built-up structural unit.

Boxed soffit See Eave, boxed.

Braced wall line Each exterior surface of a residence.

Braced wall panel A method of reinforcing a braced wall line using panels 48 inches (1,200 mm) wide to resist lateral loads.

Branch lines Water feeder lines that branch off the main line to supply fresh water to fixture groups in the home.

Breaker An electrical safety switch that automatically opens the circuit when excessive amperage occurs in the circuit.

Breezeway A covered walkway with open sides between two different parts of a structure.

Bridging Cross blocking between horizontal members used to add stiffness. Also called blocking.

Building code Legal requirements designed to protect the public by providing guidelines for structural, electrical, plumbing, and mechanical areas of a structure. (See International Residential Code.)

Building envelope The portion of a building that encloses the treated environment including the walls, ceiling or roof, and floor.

Building line An imaginary line determined by zoning departments to specify the limits where a structure may be built (also known as a setback).

Building paper A waterproofed paper used to prevent the passage of air and water into a structure.

Building permit A permit to build a structure issued by a governmental agency after the plans for the structure have been examined and the structure is found to comply with all building code requirements.

Built-up beam A beam built to smaller members that are bolted or nailed together.

Built-up roof A roof composed of three or more layers of felt, asphalt, pitch, or coal tar.

Bullnose Rounded edges of cabinet trim.

Butt joint The junction where two members meet in a square-cut joint; end to end, or edge to edge.

Buttress A projection from a wall often located below roof beams to provide support to the roof loads and to keep long walls in the vertical position.

Cabinet work The interior finish woodwork of a structure, especially cabinetry.

Camber A curvature that is built into a laminated beam to increase its load-bearing ability.

Cant strip A small built-up area between two intersecting roof shapes to divert water.

Cantilever Projected construction that is fastened at only one end.

Carport A covered automobile parking structure that is not fully enclosed.

Carriage The horizontal part of a stair stringer that supports the tread.

Casement window A hinged window that swings outward around a vertical axis.

Casing The metal, plastic, or wood trim around a door or a window.

Catch basin An underground reservoir for water drained from a roof before it flows to a storm drain.

Cathedral window A window with an upper edge that is parallel to the roof pitch.

Caulking A soft, waterproof material used to seal seams and cracks in construction.

Cavity wall A masonry wall formed with two wythes with an airspace between each face.

Ceiling joist The horizontal member of the roof that is used to resist the outward spread of the rafters and to provide a surface to mount the finished ceiling.

Cement A powder of alumina, silica, lime, iron oxide, and magnesia pulverized and used as an ingredient in mortar and concrete.

Central heating A heating system that delivers heat throughout a structure from a single source.

Cesspool An underground catch basin for the collection and dispersal of sewage.

Chair rail Molding placed horizontally on the wall at the height where chair backs would otherwise damage the wall.

Chamfer A beveled edge formed by removing the sharp corner of a piece of material.

Channel A standard form of structural steel with three sides at right angles to each other forming the letter C.

Chase A recessed area of column formed between structural members for electrical, mechanical, or plumbing materials.

Check Lengthwise cracks in a board caused by natural drying.

Check valve A valve in a pipe that permits flow in only one direction.

Chimney An upright structure connected to a fireplace or furnace that passes smoke and gases to outside air.

Chimney cap The sloping surface on the top of the chimney.

Chimney hood A covering that is placed over the flue to keep the elements from entering the flue.

Chimney liner A fire clay or terra-cotta liner built into a chimney to provide a smooth surface to the chimney flue.

Chord The upper and lower members of a truss that are supported by the web.

Cinder block A block made of cinder and cement used in construction.

Circuit The various conductors, connections, and devices found in the path of electrical flow from the source through the components and back to the source.

Circuit breaker A safety device that opens and closes an electrical circuit.

Clapboard A tapered board used for siding that overlaps the board below it.

Cleanout A fitting with a removable plug that is placed in plumbing drainage lines to allow access for cleaning out the pipe.

Clearance A clear space between building materials to allow for airflow or access.

Clerestory A window or group of windows that are placed above the normal window height, often between two roof levels.

Code A performance-based description of the desired results with wide latitude allowed to achieve the results.

Coffered ceiling A ceiling formed using beams and trim to create a pattern of recessed panels or grid-like compartments in a ceiling.

Collar ties Horizontal ties between rafters near the ridge to help resist the tendency of the rafters to separate.

Colonial A style of architecture and furniture adapted from the American colonial period.

Column A vertical structural support, usually made of steel.

Common wall The partition that divides two different dwelling units.

Complex beam Beam with a non-uniform load at any point on it and has supports that are not located at its end.

Compression A force that crushes or compacts.

Concentrated load A load centralized in a small area. The weight supported by a post results in a concentrated load.

Concrete A building material made from cement, sand, gravel, and water.

Concrete blocks Blocks of concrete that are precast. The standard size is 8 × 8 × 16 inches (200 × 200 × 400 mm).

Condensation The formation of water on a surface when warm air comes in contact with a cold surface.

Conditional use A use for a property that is not allowed outright by zoning regulations, but may be allowed on a case-by-case basis, if certain conditions are met.

Conductor Any material that permits the flow of electricity.

Conduit A bendable metal, fiber pipe, or tube used to enclose one or more electrical wires.

Construction joint A joint used when concrete construction must be interrupted and is used to provide a clean surface when work can be resumed.

Continuous beam A single beam that is supported by more than two supports.

Contours A line that represents land formations.

Contractor The manager of a construction project, or one specific phase of it.

Control joint (construction joint) An expansion joint in a masonry wall formed by raking mortar from the vertical joint.

Control point survey A survey method that establishes elevations that is recorded on a map.

Convenience outlet An electrical receptacle that allows current to be drawn for an appliance.

Coping A masonry cap placed on top of a block or brick wall to protect it from water penetration.

Corbel A ledge formed in a wall by building out successive courses of masonry.

Cornice The part of the roof that extends out from the wall. Sometimes referred to as the eave.

Counterflash A metal flashing used under normal flashing to provide a waterproof seam.

Course A continuous row of building material such as shingles, stone, or brick.

Court An exterior space that is at grade level, enclosed on three or more sides by walls or a building that is open and unobstructed to the sky.

Cove lighting Lighting concealed behind a cornice or other ceiling features that directs the light upward.

Crawl space The area between the floor joists and the ground.

Cricket A diverter built to direct water away from an area of a roof where it would otherwise collect, such as behind a chimney.

Cripple A wall stud that is cut at less than full length (also referred to as a Jack stud).

Cross bracing Boards fastened diagonally between structural members, such as floor joists, to provide rigidity.

Cul-de-sac A dead-end street with no outlet that provides a circular turn-around.

Culvert Underground passageways for water, usually part of a drainage system.

Cupola A short windowed tower, or dome, typically located in the center of a flat or low-slope roof on traditional homes to provide light or ventilation.

Cure The process of concrete drying to its maximum design strength, usually taking 28 days.

Cut material Soil that is removed so that the original ground elevation is lowered.

Damper A moveable plate that controls the amount of draft for a woodstove, fireplace, or furnace.

Datum A reference point for starting a survey.

Daylight The point represented on a grading plan that represents the intersection between cut and fill.

Dead load The weight of building materials or other immovable objects in a structure.

Deadening board A material used to control the transmission of sound.

Deck An exterior floor supported on at least two opposing sides by adjoining structures, posts, or piers.

Decking A wood material used to form the floor or roof, typically used in 1- and 2-inch thicknesses.

Density The number of people allowed to live in a specific area of land or to work in a specific area of a structure.

Dentil Molding made from a series of closely spaced, rectangular blocks. It is typically found below the cornice along the roofline of a building but it can be used as a decorative band anywhere on the structure.

Details Enlargements of specific areas of a structure that are drawn where several components intersect or where small members are required.

Diaphragm A rigid plate that acts similar to a beam and can be found in the roof level, walls, or floor system.

Diffusers The outlets that supply treated air from the HVAC system into a room.

Dimension line A line that extends between two extension lines to show the length of a specific feature.

Dimension lumber Lumber ranging in thickness from 2 to 4 inches (50 × 100 mm) and having moisture content of less than 19 percent.

Distribution panel Panel where the conductor from the meter base is connected to individual circuit breakers, that are connected to separate circuits for distribution to various locations throughout the structure.

Diverter A metal strip used to divert water.

Dormer A structure that projects from a sloping roof to form another roofed area. This new area is typically used to provide a surface to install a window.

Double glazing Glazing in a door or window that is constructed from two layers of glazing.

Double hung A type of window that allows the upper and lower halves to slide past each other to provide an opening at the top and bottom of the window.

Double wall construction A method of construction used in cold climates that places the exterior finishing material over sheathing that is placed over a water-resistant membrane placed over the wall studs.

Downspout A pipe that carries rainwater from the gutters of the roof to the ground.

Drain A collector for a pipe that carries waste water from each plumbing fixture to the waste line of the building drainage system.

Drainage grate The grate is a metal cover that allows water to flow into a catch basin without allowing anyone to fall in. Water flows through the grate, into a catch basin, and then is funneled into pipes that connect to public storm sewers.

Dressed lumber Lumber that has been surfaced by a planing machine to give the wood a smooth finish.

Dry rot A type of wood decay caused by fungi that leaves the wood a soft powder.

Dry well A shallow well used to disperse water from the gutter system.

Drywall An interior wall covering installed in large sheets made from gypsum board.

Ducts Pipes, typically made of sheet metal, used to conduct hot or cold air of the HVAC system.

Duplex outlet A standard electrical convenience outlet with two receptacles.

Dutch door A type of door that is divided horizontally in the center so that each half of the door may be opened separately.

Dutch hip A type of roof shape that combines features of a gable and a hip roof.

Dynamic loads The loads imposed on a structure from a sudden gust of wind or from an earthquake.

Easement A right to make limited use of another's real property that is recorded on the deed, and survives any sale of the property for use as a public right of way, such as a utility easement that grants access to private land to place or maintain a utility.

Eave The lower part of the roof that projects from the wall (also see Cornice).

Eave, boxed An eave with a covering applied directly to the bottom side of the rafter or truss tails.

Effluent Treated sewage from a sewage treatment plan or septic tank.

Egress A term used in building codes to describe access.

Elbow An L-shaped plumbing pipe.

Elevation The height of a specific point in relation to another point. The exterior views of a structure.

Eminent domain The right of a government to condemn private property so that it may be obtained for public use.

Enamel A paint that produces a hard, glossy, smooth finish.

Engineered lumber Structural components made by turning small pieces of wood into framing members such as joists, studs, and rafters.

Entourage The surroundings of a rendered building consisting of ground cover, trees, people, and automobiles.

Equity The value of real estate in excess of the balance owed on the mortgage.

Ergonomics The study of human space and movement needs as they relate to a given work area, such as a kitchen.

Excavation The removal of soil for construction purposes.

Expansion joint A joint installed in concrete construction to reduce cracking and to provide workable areas.

Extension line Line showing the extent of a dimension that are generally thin, dark, crisp lines.

Fabrication Work done on a structure away from the job site.

Facade The exterior covering of a structure.

Face brick Brick that is used on the visible surface to cover other masonry products.

Face grain The pattern in the visible veneer of plywood.

Fanlight A semi-circular or semi-elliptical, non-opening transom window with a horizontal sill placed above a door or another window that is typically found above the main entry door of Federal houses.

Fascia A horizontal board nailed to the end of rafters or trusses to conceal their ends.

Federal Housing Administration (FHA) A governmental agency that insures home loans made by private lending institutions.

Felt A tar-impregnated paper used for water protection under roofing and siding materials. Sometimes used under concrete slabs for moisture resistance.

Fenestration Windows or doors located in the building envelope.

Fiber bending stress (F_b) The measurement of structural members used to determine their stiffness.

Fiberboard Fibrous wood products that have been pressed into a sheet. Typically used for the interior construction of cabinets and for a covering for the subfloor.

Fill Material used to raise an area for construction. Typically gravel or sand is used to provide a raised, level building area.

Filled insulation Insulation material that is blown or poured into place in attics and walls.

Finished grade The shape of the ground once all excavation and movement of earth has been completed.

Finished lumber Wood that has been milled with a smooth finish suitable for use as trim and other finish work.

Finished size Sometimes called the dressed size, the finished size represents the actual size of lumber after all milling operations and is typically about ½ inch (13 mm) smaller than the nominal size, which is the size of lumber before planing.

Nominal Size (inches)	Finished Size (inches)
1	¾
2	1½
4	3½
6	5½
8	7½
10	9½
12	11½
14	13½

Fire cut An angular cut on the end of a joist or rafter that is supported by masonry. The cut allows the wood member to fall away from the wall without damaging a masonry wall when wood is damaged by fire.

Fire door A door used between different types of construction that has been rated as being able to withstand fire for a certain amount of time.

Fire rated A rating given to building materials to specify the amount of time the material can resist damage caused by fire.

Fire wall A wall constructed of materials resulting in a specified time that the wall can resist fire before structural damage will occur.

Firebox The combustion chamber of the fireplace where the fire occurs.

Firebrick A refractory brick capable of withstanding high temperatures and used for lining fireplaces and furnaces.

Fireplace insert A metal fireplace that is inserted into a masonry fireplace to control drafts and increase heat production. The unit must be vented using the existing chimney.

Fireplace opening The open area between the side and top faces of the fireplace.

Fireproofing Any material that is used to cover structural materials to increase their fire rating.

Fire-stop Blocking placed between studs or other structural members to resist the spread of fire.

Fitting A standard pipe or tubing joint such as a tee, elbow, or reducer used to join two or more pipes.

Fixed window Window that is designed without hinges so it cannot be opened.

Flagstone Flat stones used typically for floor and wall coverings.

Flashing Metal used to prevent water from leaking through surface intersections.

Flat roof A roof with a minimal roof pitch, usually about ¼ inch per 12 inches (6 per 25).

Flitch beam A built-up beam consisting of steel plates bolted between wood members.

Floor joists Repetitive horizontal structural members of the floor framing system that are used to span between the stem wall or girders to provide support to the subfloor.

Floor plug A 120-volt convenience outlet located in the floor.

Flue A passage inside of the chimney to conduct smoke and gases away from a firebox to outside air.

Flue liner A terra-cotta pipe used to provide a smooth flue surface so that unburned materials will not cling to the flue.

Footing The lowest member of a foundation system used to spread the loads of a structure across supporting soil.

Forced-air heating systems A system that blows treated air through ducts by use of a fan located in the heating or cooling device.

Formal balance Symmetrical arrangement of space so that one side of the structure or room matches the opposite side in size.

Foundation The system used to support a building's loads and made up to stem walls, footings, and piers. The term is used in many areas to refer to the footing.

Frame The structural skeleton of a building.

French doors Exterior or interior doors that have glass panels and swing into a room.

Frieze A horizontal band decorated with designs or carvings that runs above doorways or windows, or below the cornice.

Frost line The average depth that soil will freeze.

Furring Wood strips attached to structural members that are used to provide a level surface for finishing materials when different sized structural members are used.

Gable A type of roof with two sloping surfaces that intersect at the ridge of the structure.

Gable end wall The triangular wall that is formed at each end of a gable roof between the top plate of the wall and the rafters.

Gambrel A type of roof formed with two planes on each side of the ridge. The lower pitch is steeper than the upper portion of the roof.

Geothermal system Heating and cooling system that uses the constant, moderate temperature of the ground to provide space heating and cooling, or domestic hot water, by placing a heat exchanger in the ground, or in wells, lakes, rivers, or streams.

Girder A horizontal support beam used at the foundation level to support the floor joists. In a post and beam system, a girder is used to support the floor decking.

Glazing All areas that let in natural light, including windows, clerestories, skylights, glass doors, glass block walls, and glass portions of doors.

Glued-laminated beam (Glu-lam) A structural member made up of layers of lumber that are glued together.

Grade The designation of the quality of a manufactured piece of wood.

Grading The moving of soil to effect the elevation of land at a construction site.

Grading plan A drawing used to show the finished soil configuration of the building site.

Gravel stop A metal strip used to retain gravel at the edge of built-up roofs.

Gravity Uniform force that affects all structures due to the gravitational force from the earth.

Gray water Wastewater from a shower, bath, laundry water, and rain runoff collected from roof gutters that is recycled.

Green lumber Lumber that has not been kiln-dried and still contains moisture.

Green board A type of water-resistant gypsum board designed to be used in high moisture areas such as behind a shower enclosure.

Ground An electrical connection to the earth by means of a rod.

Ground fault circuit interrupter (GFCI or GFI) A 120-convenience outlet with a built-in circuit breaker that must be used within 60 inches of any water source.

Grout A mixture of cement, sand, and water used to fill joints in masonry and tile construction.

Guardrail A horizontal protective railing used around stairwells, balconies, and changes of floor elevation greater than 30 inches.

Gusset A metal or wood plate used to strengthen the intersection of structural members.

Gutter A device mounted on an eave for the collection of rainwater from the roof to downspouts.

Gypsum board An interior finishing material made of gypsum and fiberglass and covered with paper that is installed in large shapes.

Habitable space Rooms used for sleeping, living, cooking, or dining purposes.

Half-timber A frame construction method where spaces between wood members are filled with masonry.

Hanger A metal support bracket used to attach two structural members.

Hardboard Sheet material formed of compressed wood fibers.

Head The upper portion of a door or window frame.

Header A horizontal structural member used to support other structural members over openings, such as doors and windows.

Header course A horizontal masonry course with the end of each masonry unit exposed.

Headroom The vertical clearance over a stairway.

Hearth The fire-resistant floor extending in front of, and to the side of the firebox.

Heartwood The inner core of a tree trunk.

Heat pump A unit designed to produce forced air for heating and cooling.

Hip A traditional roof shape formed by four or more inserting planes. The term is also used to describe an exterior edge formed by two sloping roof surfaces.

Hip roof A roof shape with four sloping sides.

Hopper window Window that is hinged at the bottom and swings inward.

Horizon line A line drawn parallel to the ground line that represents the intersection of ground and sky.

Horizontal shear (F_v) One of three major forces acting on a beam, it is the tendency of the fibers of a beam to slide past each other in horizontal direction.

Hose bibb A water outlet that is threaded to receive a hose.

Hue A term used to represent what you typically think of as the color.

Humidifier A mechanical device that controls the amount of moisture inside of a structure.

Hurricane ties Metal connectors used to connect roof members to wall members to resist uplift.

Hydroelectric power Electricity generated by the conversion of the energy created by falling water.

I-beam The generic term for a wide flange or American standard steel beam with a cross section in the shape of the letter *I*.

Indirect lighting Mechanical lighting that is reflected off a surface.

Infiltration The flow of air through building intersections.

Informal balance Nonsymmetrical placement achieved by placing shapes of different sizes in various positions around the imaginary centerline.

Insulated concrete form (ICF) An energy-efficient wall framing system; poured concrete is placed in polystyrene forms that are left in place to create a super-insulated wall.

Insulation Material used to restrict the flow of heat, cold, or sound from one surface to another.

Interior decorator A person who decorates the interiors of buildings, with the aim of making rooms more attractive, comfortable, and functional.

International Residential Code (IRC) A national building code for one- and two-family dwellings.

Intensity The brightness or strength of a specific color.

Interpolation A combination of rounding off and guessing based on known points. If you know that two points have a change of elevation of 12 inches (25 mm), you can identify a point halfway between these two points and assign it an elevation representing a 6-inch (13-mm) difference in height.

Irrigation plan A drawing, usually completed by a technician working for a landscape architect, that shows how landscaping will be maintained.

Isolation joint See expansion joint.

Isometric drawings Drawings that appear to be three-dimensional drawings by showing three surfaces of an object in one view.

Jack rafter A rafter that is cut shorter than the other rafters to allow for an opening in the roof.

Jack stud (Cripple) A wall member that is cut shorter than other studs to allow for an opening such as a window.

Jalousie A type of window made of thin horizontal panels that can be rotated between the open and closed position.

Jamb The vertical members of a door or window frame.

Joist A horizontal structural member used in repetitive patterns to support floor and ceiling loads.

Junction box A box that protects electrical wiring splices in conductors or joints in runs.

Kick block (Kicker) A block used to keep the bottom of the stringer from sliding on the floor when downward pressure is applied to the stringer.

Kiln dried A method of drying lumber in a kiln or oven. Kiln-dried lumber has a reduced moisture content when compared to lumber that has been air dried.

King stud A full-length stud placed at the end of a header and beside the trimmer stud that is used to support the header.

Kip Used in some engineering formulas to represent 1,000 lb.

Knee wall A wall of less than full height.

Knot A branch or limb of a tree that is cut through in the process of manufacturing lumber.

Lag screw Screws used for wood-to-wood or wood-to-steel connections.

Lally column A vertical steel column that is used to support floor or foundation loads.

Laminated Several layers of material that have been glued together under pressure.

Landing A platform between two flights of stairs.

Landscape plan A drawing that shows the location, type, size and quantity of all vegetation required for the project as well as hardscaping such as patios, walkways, fountains, pools, sports courts, and other landscaping features.

Lateral Sideways motion in a structure caused by wind or seismic forces. The term is also used to describe the pipe that connects the construction site to the public sewer pipe.

Lath Wood or sheet metal strips that are attached to the structural frame to support plaster.

Lattice A grille made by crisscrossing strips of material.

Lavatory A bathroom sink, or a room that is equipped with a washbasin.

Leach lines Soil-absorbent field used to disperse liquid material from a septic system.

Ledger A horizontal member that is attached to the side of wall members to provide support for rafters or joists.

Legal description The description used to describe a parcel of land for legal purposes such as the recording of a deed of ownership.

Lintel A horizontal steel member used to provide support for masonry over an opening.

Lintel block A long, rectangular stone block that spans a door or window opening to support the weight of the structure above the opening.

Lisp A programming language used to customize CAD software.

Live load The load from all movable objects within a structure including loads from furniture and people. External loads from snow and wind are also considered live loaded.

Load-bearing wall A support wall that holds floor or roof loads in addition to its own weight.

Load path The route that is used to transfer the roof loads into the walls, then into the floor, and then to the foundation.

Lookout Bracing between the wall and sub-fascia or end cap that the soffit attaches to. Also a beam used to support eave loads.

Louver An opening with horizontal slats to allow for ventilation.

Low-e glass Low-emission glass that has a transparent coating on its surface that acts as a thermal mirror.

Main The water supply line that extends from the water meter into the home to deliver potable water.

Manifold A distribution center between the main and branch and riser lines.

Mansard A four-sided, steep-sloped roof used to enclose the upper level of a structure.

Mantel A decorative shelf above the opening of a fireplace.

Market value The amount that property can be sold for.

Masonry The use of brick, stone, or concrete blocks to construct a wall.

MasterFormat™ A list of numbers and titles that are created to organize information into a standard sequence that relates to construction requirements, products, and activities.

Mesh A metal reinforcing material placed in concrete slabs and masonry walls to help resist cracking.

Metal ties A manufactured piece of metal for joining two structural members together.

Metal wall ties Corrugated metal strips used to bond brick veneer to its support wall.

Metric measurement—hard conversions Made by using a mathematical formula to change a value of one system (e.g. 1 inch) to the equivalent value in another system (e.g. 25.4 mm). A 6-inch distance would be a 152 mm (6 × 25.4).

Metric measurement—soft conversions Made by using a mathematical formula to change a value from one system (e.g. 1 inch) to a rounded value in another system (e.g. 25 mm). A 6-inch distance would be a 150 mm.

Millwork Finished woodwork that has been manufactured in a milling plant. Examples are window and door frames, mantels, moldings, and stairway components.

Mineral wool An insulating material made of fibrous foam.

Modular cabinet Prefabricated cabinets that are constructed in specific sizes called modules. Modular cabinets are usually available in three inch (75 mm) widths.

Module A standardized unit of measurement.

Moisture barrier Typically a plastic material used to restrict moisture vapor from penetrating into a structure.

Molding Decorative strips, usually made of wood, used to conceal the seam in other finishing materials.

Moment The tendency of a force to rotate around a certain point.

Monolithic Concrete construction created in one pour.

Monument A point established by the U.S. Geological Society (USGS) that is marked by a steel rod or a benchmark. A monument is referred to as the true point of beginning in a metes and bounds legal description.

Mortar A combination of cement, sand, and water used to bond masonry units together.

Moving loads Loads that are not stationary such as those produced by automobiles and construction equipment.

Mudroom A room or utility entrance where soiled clothing can be removed before entering the main portion of the residence.

Mudsill (Base plate) The horizontal wood member that rests on concrete to support other wood members.

Mullion A horizontal or vertical divider between sections of a window.

Muntin A horizontal or vertical divider within a section of a window.

Nailer A wood member bolted to concrete or steel members to provide a nailing surface for attaching other wood members.

National Council for Interior Design Qualification (NCIDQ) The board that regulates the standards to become a professional interior designer.

Natural Grade Soil in its unaltered state.

Nested joists The practice of placing a steel joist around another joist so that the strength of the joist is doubled.

Net size Final size of wood after planing.

Neutral axis The axis formed where the forces of compression and tension in a beam reach equilibrium.

Newel The end post of a stair railing.

Nominal size An approximate size achieved by rounding the actual material size to the nearest larger whole number.

Nonferrous metal Metal, such as copper or brass, that contains no iron.

Non-habitable spaces Include closets, pantries, bath or toilet rooms, hallways, utility rooms, storage spaces, garages, darkrooms, and other similar spaces.

Non-bearing wall A wall that supports no loads other than its own.

Nosing The rounded front edge of a tread that extends past the riser.

Obscure glass Glass that is not transparent.

On center A measurement taken from the center of one member to the center of another member.

Oriel window A window placed in a projection of an exterior wall that does not extend all the way to the foundation. In plan view, the wall projection may be rectangular, polygonal, or curved. See Bay window.

Orientation The locating of a structure on property based on the location of the sun, prevailing winds, view, and noise.

Oriented Strand Board (OSB) A sheet of material made of layers of wood chips laminated together with glue under extreme pressure. The standard size is a 4 × 8-foot sheet and it is typically used for the same applications as plywood.

Outlet An electrical receptacle that allows for current to be drawn from the system.

Outrigger A support for roof sheathing and the fascia that extends past the wall line perpendicular to the rafters.

Overhang The horizontal measurement of the distance the roof projects from a wall.

Palladian window A large window divided into three parts, typical of classical architectural styles such as Georgian and Federal. The arched center section is larger than the two rectangular side sections. A typical location for a Palladian window is above the front door.

Parapet A portion of wall that extends above the edge of the roof.

Parging A thin coat of plaster used to smooth a masonry surface.

Parquet flooring Wood flooring laid to form patterns.

Partition An interior wall.

Party wall A wall dividing two adjoining spaces such as apartments or offices.

Passive solar system System that uses natural, architectural means to store and radiate solar heat.

Patio Ground-level exterior entertaining area that is made of concrete, stone, brick, or treated wood.

Pediment A low-pitched triangular gable based on Greek Revival style of architecture placed over a door or window, or on the front of buildings.

Penny The length of a nail represented by the lowercase letter *d*; "10d" is read as "ten penny."

Percolation test Test used to determine if the soil could accommodate a septic system.

Permit plans Plans required by a government agency for permission to build a structure.

Pier A concrete or masonry foundation support.

Pilaster A reinforcing column built into or against a masonry wall.

Piling A vertical foundation support driven into the ground to provide support on stable soil or rock.

Pitch A description of roof angle comparing the vertical rise to the horizontal run.

Plank Lumber that is 1½ to 2½ inches in thickness.

Plaster A mix of sand, cement, and water used to cover walls and ceilings.

Plat A map of an area of land that shows the boundaries of individual lots.

Plate Horizontal pieces of wood used at the top and bottom of a wall to keep the studs in position.

Platform framing (Western platform) A building construction method where each floor acts as a platform in the framing.

Plenum An enclosed air space for transporting air from the HVAC system. The air pressure in the plenum is greater than the pressure in the structure causing air to flow from the furnace through the plenum and into the residence.

Plot A parcel of land. Also used to make a paper copy of a CAD drawing.

Plumb True vertical.

Plywood Wood composed of three or more layers with the grain of each layer placed at 90°F to each other and bonded with glue.

Pocket door A door that slides into a pocket that has been built into the wall.

Point-of-beginning Fixed location on a plot of land where the survey begins.

Porch A roofed entrance to a structure that is that open to the air and supported on one side by the home with the remaining sides supported by columns or arches.

Portal frame (PF) A method of reinforcing a braced wall line that uses two panels with a minimum width of 22½ inches (570 mm). The panels are connected by a header to resist lateral loads.

Porte-cochere A porch large enough for wheeled vehicles to pass through.

Portico A roofed area that is open to the air on one or more sides, supported on one side by the home and by columns or arches on the remaining sides. Porticos are common on Federal, Early Classical Revival, and Greek Revival homes.

Portland cement A hydraulic cement made of silica, lime, and aluminum that has become the most common cement used in the construction industry because of its strength.

Post A vertical wood structural member usually 4 × 4 or larger.

Post-tensioning Concrete slabs that are poured over unstable soil by using a method of reinforcement.

Power driven steel studs Smooth bolt-like fasteners driven by a powder actuated fastening tool to fasten wood members to concrete or steel members.

Precast A concrete component that has been cast in a location other than where it will be used.

Prefabricated Buildings or components that are built away from the job site and transported ready to be used.

Prestressed A concrete component that is placed in compression as it is cast to help resist deflection.

Prevailing winds Direction from which the wind most frequently blows in a given area of the country.

Profile Vertical section of the surface of the ground and/or of underlying earth that is taken along any desired fixed line.

Proportion A pleasing relationship related to both size and balance. Rectangles using the proportions 2:3, 3:5, and 5:8 have long been considered to be pleasing.

Punch-out A hole in the web of a steel framing member allowing for the installation of plumbing, electrical, and other trade installation

Purlin A horizontal roof member that is laid perpendicular to rafters to help limit deflection.

Purlin brace A support member that extends from the purlin down to a load-bearing wall or header.

Pyramidal roof A hipped roof that lacks a ridge. The four isosceles-triangular planes of the roof meet at a common apex that resembles a pyramid. Low-slope pyramidal roofs are common on Greek Revival houses.

Quad A courtyard surrounded by the walls of buildings.

Quarry tile An unglazed, machine-made tile.

Quarter round Wood molding that has the profile of one-quarter of a circle.

Quatrefoil window A round window common in Moorish, Gothic, and Mission style architecture. The window is composed of four equal lobes, like a four-petaled flower.

Quoin Heavy blocks of stone found at corners of a brick building that are designed to reinforce masonry walls. They are often found on Georgian and some Federal and Greek Revival houses. In a brick house, the quoins usually consist of granite blocks, but may also be formed from bricks and painted in the trim color. Wood quoins are made to imitate stone and are strictly decorative ornamental corner features.

Rabbet A rectangular groove cut on the edge of a board.

Radiant barrier Material made of aluminum foil with backing; used to stop heat from radiating through the attic.

Radiant heat Heat emitted from a particular material such as brick, electric coils, or hot water pipe without use of air movement.

Radon A naturally occurring radioactive gas, usually found in a basement, that breaks down into compounds that are carcinogenic when inhaled over long periods of time.

Rafter The inclined structural member of a roof system designed to support roof loads.

Rafter/ceiling joist An inclined structural member that supports both the ceiling and the roof materials.

Rafter, common A rafter that extends the full length between the support wall and the ridge board.

Rafter, hip A rafter used at the intersection of two roof planes that forms a sloping ridge (an exterior roof corner).

Rafter, jack A rafter that spans from the supporting wall to a hip or valley rafter so that the jack rafter is not the full length of a common rafter.

Rafter, valley A rafter placed at the intersection of two roof planes that forms an interior roof intersection.

Rake Roof extension projecting over an endwall following the slope of the roof.

Rake joint A recessed mortar joint.

Rebar Reinforcing steel used to strengthen concrete.

Reference bubble A symbol used to designate the origin of details and sections.

Register An opening in a duct for the supply of heated or cooled air.

Reinforced concrete Concrete that has steel rebar placed in it to resist tension.

Relative humidity The amount of water vapor in the atmosphere compared to the maximum possible amount at the same temperature.

Remodel A construction project that involves moving, adding, or removing structural members.

Renovation The altering or removing and replacing of non-structural materials such as cabinets, and minor electrical or mechanical repairs.

Retaining wall A wall, usually made of masonry, that is designed to resist soil loads.

R-factor A unit of thermal resistance applied to the insulating value of a specific building material.

Rheostat An electrical control device used to regulate the current reaching a light fixture. A dimmer switch.

Ribbon A structural wood member framed into studs to support joists or rafters.

Ridge The uppermost area of two intersecting roof planes.

Ridge beam A beam located at the ridge used to support the roof framing members. A beam can also be used as a decorative member at an interior ridge.

Ridge board The horizontal member at the ridge that runs perpendicular to the rafters. The rafters are aligned against the ridge to resist the downward force of the rafters.

Ridge brace A support member used to transfer the weight from the ridge board to a bearing wall or beam. The brace is typically spaced at 48 inches O.C. and may not exceed a 45°F angle from vertical.

Rim joist A joist at the perimeter of a structure that runs parallel to the other floor joist. Sometimes referred to as a band or header.

Rim track A metal track installed over the end of steel floor joists to support the ends and close off the space between them.

Rise The amount of vertical distance between one tread and another. Also refers to the angle of the roof based on 12 horizontal units.

Riser The vertical member of stairs between the treads. The term is also used to refer to a water supply pipe that extends vertically one or more stories to carry water to fixtures.

Roll roofing Roofing material of fiber or asphalt that is shipped in rolls.

Roof drain A receptacle for removal of roof water.

Rough floor The subfloor, usually plywood, that serves as a base for the finished floor.

Rough-in To add framing for a feature such as a door or window that will be installed at a future date. The opening is covered with the finish materials until the door or window is needed. The term is also used to describe the HVAC, plumbing, or electrical work that is done in the joist or stud space prior to adding the finish material.

Rough lumber Lumber that has not been surfaced but has been trimmed on all four sides.

Rough opening The unfinished opening provided between framing members to allow for the placement of doors, windows, or other skylights

Rowlock A pattern for laying masonry units so that the end of the unit is exposed.

Run The horizontal distance of a set of steps or the measurement describing the depth of one step. Also used to describe the horizontal measurement from the outside edge of the wall to the centerline of the ridge.

R-value Measurement of thermal resistance used to indicate the effectiveness of insulation.

Rhythm A repetitive element that leads the eye through the design from one place to another in an orderly fashion.

Saddle A small gable-shaped roof used to divert water from behind a chimney.

Sanitary sewer A sewer line that carries sewage without any storm, surface, or groundwater.

Sash An individual frame around a window.

Sawn lumber Lumber that comes from trees without being altered by engineering.

Scab A short member that overlaps the butt joint of two other members used to fasten those members.

Scale A ratio that is used to reduce or enlarge the size of a drawing for plotting. *Scale* is also used to refer to a measuring tool.

Schedule A written list of similar components such as windows and doors.

Sconce A wall-mounted light fixture that provides indirect lighting by directing light either up or down, depending on the shape of the sconce.

Scratch coat The first coat of stucco that is scratched to provide a good bonding surface for the second coat.

Seasoning The process of removing moisture from green lumber by either air (natural) or kiln drying.

Section A type of drawing showing an object as if it had been cut through to show interior construction.

Seismic Earthquake-related forces.

Self-drilling screws A fastener with a drilling point that is able to penetrate heavy-gauge metal.

Septic system A sewage disposal system consisting of a storage tank and an absorption field. Sewage is stored in the septic tank and the liquid waste is dispersed into the drainage field.

Septic tank A tank where sewage is decomposed by bacteria and dispersed by drain tiles.

Service connection The wires that run to a structure from a power pole or transformer.

Setback The minimum distance required between the structure and the property line.

Shake A hand-split wooden roof shingle.

Shear The stress that occurs when two forces from opposite directions are acting on the same member. Shearing stress tends to cut a member just as scissors cut paper.

Shear panel A wall panel designed by an engineer to resist wind or seismic forces.

Sheathing A thin covering that is usually made of OSB or plywood with a thickness of between $3/8$ and $3/4$ inch (9.5 or 19 mm) that is placed over walls, floors, and roofs. It serves as a backing for the finish materials.

Shim A piece of material used to fill a space between two surfaces.

Shiplap A siding pattern of overlapping rabbeted edges.

Sidelight A window or a series of small fixed panes arranged vertically, found on either side of the main entry door of many Federal, Greek Revival homes.

Sill A horizontal wood member placed at the bottom of walls and openings in walls.

Simple beam Beam with a uniform load evenly distributed over its entire length and supported at each end.

Single wall construction A construction method used in temperate climates that places the exterior finishing material directly over a water-resistant membrane and the wall studs.

Site orientation Placement of a structure on a property with certain environmental and physical factors taken into consideration.

Skip sheathing Material such as 1×4s (25×100s) that are laid perpendicular to the rafters to support roofing such as wood shingles and concrete or clay tiles.

Skylight An opening in the roof to allow light and ventilation, usually covered with glass or plastic.

Sky window A opening in the wall and roof that combines features of a skylight and a window using glazing on a wall that extends to meet glazing on a portion of the roof.

Slab A concrete floor system typically poured at ground level.

Sleepers Strips of wood placed over a concrete slab in order to attach other wood members.

Smoke chamber The portion of the chimney located directly over the firebox that acts as a funnel between the firebox and the chimney.

Smoke shelf A shelf located at the bottom of the smoke chamber to prevent down-drafts from the chimney from entering the firebox.

Soffit A lowered ceiling, typically found in kitchens, halls, and bathrooms to allow for recessed lighting or HVAC ducts. The term is also used to describe an enclosed area below the overhangs that is used to protect the rafter or truss tails from the elements.

Softwood Wood that comes from cone-bearing trees.

Soil line Disposal lines in the waste water system.

Soil stack The main vertical waste water pipe.

Soil vent A vent that runs up the wall and vents out the roof, allowing vapor to escape and ventilating the system.

Solar heat Heat that comes from energy generated from sunlight.

Solar panel Solar cells that convert sunlight into electricity.

Solarium A glassed-in porch on the south side of a house that is in direct exposure to the sun's rays.

Soldier A masonry unit laid on end with its narrow surface exposed.

Sole plate The plate placed at the bottom of a wall.

Spackle The covering of sheetrock joints with joint compound.

Span The horizontal distance between two supporting members.

Spark arrester A screen placed at the top of the flue to prevent combustibles from leaving the flue.

Specifications A written statement that describes the characteristics of a particular aspect of a project, such as materials, or equipment, construction systems standards, and workmanship.

Splice Two similar members that are joined together in a straight line, usually by nailing or bolting.

Split-level A house that has two levels, one about half a level above or below the other.

Spot grade An contour elevation determined by a survey team of a specific location at the job site.

Square An area of roofing covering 100 sq ft.

Stack A vertical drain lines that carry waste from the home to the sewer main.

Stair jack See Stringer.

Stairwell The opening in the floor where a stair will be framed.

Station point The position of the observer's eye in a perspective drawing.

Stations Numbers given to the horizontal lines of a grid survey.

Stick framed Framing one member at a time on the job site, instead of raising prefabricating walls or Red Iron Frames as a unit.

Stile A vertical member of a cabinet, door, or decorative panel.

Stirrup A U-shaped metal bracket used to support wood beams.

Stock Common sizes of building materials.

Stock plans Houses designed to appeal to a wide variety of people, with several different options.

Stop A wooden strip used to hold windows in place.

Storm sewer A municipal drainage system used to dispose groundwater, rainwater, surface water, or other nonpolluting waste separately from sewage disposal.

Stress A live or dead load acting on a structural member. Stress results as the fibers of a beam resist an external force.

Stressed-skin panel A hollow, built-up member typically used as a beam.

Stretcher A course of masonry laid horizontally with the end of the unit exposed.

Stringer (Stair jack) The inclined support member of a stair that supports the risers and treads.

Strong back A beam used to support ridge and ceiling loads. It is placed above the ceiling joists when perpendicular to the joists and between the joist when it is parallel to the joists.

Stucco A type of plaster made from Portland cement, sand, water, and a coloring agent that is applied to exterior walls.

Stud The vertical repetitive framing members of a wall that are usually 2×4 (50×100) or 2×6 (50×150) in size.

Subfloor The flooring surface that is laid on the floor joint and serves as a base layer for the finished floor.

Subsill A sill located between the trimmers and bottom side of a window opening. It provides a nailing surface for interior and exterior materials.

Sump A recessed area in a basement floor to collect water so that it can be removed by a pump.

Sun tunnel A type of skylight that delivers light into a room using a reflective, flexible tunnel to connect the roof opening with an opening in the ceiling. A diffuser is mounted at the ceiling end of the tunnel to disperse the light.

Surfaced lumber Lumber that has been smoothed on at least one side.

Survey map Map of a property showing its size, boundaries, and topography.

Swale A recessed area formed in the ground to help divert ground water away from a structure.

Tamp To compact soil or concrete.

Temporary loads The loads that must be supported by a structure for a limited time.

Tensile strength The resistance of a material or beam to the tendency to stretch.

Tension Forces that cause a material to stretch or pull apart.

Termite shield A strip of sheet metal used at the intersection of concrete and wood surfaces near ground level to prevent termites from entering the wood.

Terra-cotta Hard-baked clay typically used as a liner for chimneys.

Terrain Characteristics that describe the shape of land.

Threshold The beveled member directly under a door.

Throat The narrow opening to the chimney that is just above the firebox. The throat of a chimney is where the damper is placed.

Timber Lumber with a cross-sectional size of 4 × 6 inches or larger.

Toenail Nails driven into a member at an angle.

Tongue and groove A joint where the edge of one member fits into a groove in the next member.

Top plate A horizontal structural member located on top of the studs used to hold the wall together.

Track A U-shaped member used for applications such as top and bottom track for walls and rim track for floor joists.

Traffic flow The route that people follow as they move from one area of a residence to another, using hallways or portions of a room.

Transom Originally used to denote a horizontal crossbar in a window. It later came to mean a window positioned above such a crossbar. Today the term is most commonly used for a shallow, rectangular window located immediately above a door.

Trap A U-shaped vented fitting below plumbing fixtures that provides a liquid seal to prevent the emission of sewer gases without affecting the flow of sewage or waste water.

Tray ceilings A ceiling constructed with the sides angling at approximately 45% or curving to a higher flat ceiling so that it resembles an inverted tray.

Tread The horizontal member of a stair, on which the foot is placed.

Tributary width The accumulation of loads that are directed to a structural member. It is always half the distance between the beam to be designed and the next bearing point.

Trimmer A stud that is not full height that is used to place a header. The term is also used to describe a joist or rafters that are used to frame an opening in a floor, ceiling, or roof.

Truss A prefabricated or job-built construction member formed of triangular shapes used to support roof or floor loads over long spans.

Truss clips See hurricane ties. Used to tie the roof-framing members to the wall.

Ultimate strength The unit stress within a member just before it breaks.

Underlayment Thin material, usually ⅜ or ½" (9 or 13 mm) plywood, waferboard, or hardboard, used to provide a smooth impact-resistant surface on which to install the finished flooring.

UniFormat An arrangement of construction information based on physical parts of a facility called systems and assemblies.

Unit stress The maximum permissible stress a structural member can resist without failing.

Unity A principle of design related to the common design or decorating pattern that ties a structure together.

Uplift The tendency of structural members to move upward due to wind or seismic pressure.

Value The darkening or lightening of a hue.

Valve A fitting that is used to control the flow of fluid or gas to a fixture or appliance.

Valley The internal corner formed between two intersecting roof structures.

Vapor barrier Material that is used to block the flow of water vapor into a structure. Typically 6-mil (0.006 inch) black plastic.

Variance A legal request by a property owner to allow a modification from a standard or requirement of the zoning code.

Varge rafter See barge rafter.

Vault An inclined ceiling area.

Veneer A thin outer covering or non-load-bearing masonry face material.

Vent pipe Pipes that allow air into the waste lines to allow drainage. Each plumbing fixture is connected to a vent stack.

Vent stack A vertical pipe of a plumbing system used to equalize pressure within the system and to vent sewer gases.

Ventilation The process of supplying and removing air from a structure.

Vertical shear A stress acting on a beam that causes the beam to drop between its supports.

Vestibule A small entrance or lobby.

Virtual reality A computer-simulated world that appears to be a real world.

Volt A unit of measurement of electrical force or potential that makes electricity flow through an electrical wire. For a specific load, the higher the voltage, the more electricity will flow.

Wainscot Paneling applied to the lower portion of a wall.

Wallboard Large, flat sheets of gypsum, typically ⅜, ½, or ⅝ inch (9.5, 13, or 16 mm) thick, used to finish interior walls.

Warp Variation from true shape.

Waterproof Material or a type of construction that prevents the absorption of water.

Watt A unit of power measurement based on the potential and the current. The amount of power to be used per fixture is be determined by multiplying the current (amps) by volts (potential) to determine the watts.

Weather strip A fabric or plastic material placed along the edges of doors, windows, and skylights to reduce air infiltration.

Webs Interior members of the truss that span between the top and bottom chords.

Web stiffener Additional material that is attached to the web of a steel C-section member to strengthen the member against web crippling.

Weep hole An opening in the bottom course of a masonry to allow for drainage.

Weld A method of providing a rigid connection between two or more pieces of steel.

Wetland Lowland areas such as marshes, swamps, and bogs that in their normal condition are saturated with moisture and therefore provide a natural habitat for certain wildlife.

Widow's walk A deck above the highest level of the roof from which sea captains' wives were supposed to have watched for their husbands' ships.

Working drawings The drawings that will be used to obtain a permit and build a structure.

Work triangle The relationship between the work areas formed by drawing lines from the centers of the storage, preparation, and cleaning areas. This triangle outlines the main traffic area required to prepare a meal with food taken from the refrigerator, cleaned at the sink, and cooked at the microwave or stove, and then leftovers returned to the refrigerator.

Wythe A single unit thickness of a masonry wall.

Zoning An ordinance that regulates the location, size, and type of a structure in a building zone.

Zoning regulations Limits to the uses of property and the structures that may be built by controlling the use of land, lot sizes, types of structures permitted, building heights, setbacks, and density (the ratio of land area to improvement area).

Index

A

ABS (acrylonitrile butadiene styrene) pipe, 230, 231
AC. *See* Alternating current
Accidents. *See specific types*
Acrylic latex caulk, 44, 241
Adapters, 233
Adhesive caulk, 45
Adhesives
    job sheet, 67
    overview, 43–46
Adjustable wrenches, 55, 229
Aeration, 343
Air conditioners. *See also* HVAC systems
    condensate systems, 279–280
    filter replacement, 266–267
    receptacles, 93
    through-the-wall systems, 280–281
Air filters, 276, 346, 347
Algae preventive, 344
Alkyd, 204
All purpose cleaners, 46
Alternating current (AC), 78, 87–88
Aluminum oxide sandpaper, 60
Amber, 71–72
Ammeters, 101, 105–106
Amperes (amp, A)
    definition, 75
    electrical theory, 87
    receptacles, 92–93
    switch types, 92
Angled jaw pliers, 228
Angular brushes, 205
Ants, 327
Appliances. *See also specific appliances*
    job sheets, 309, 311
    load carry capacity, 79
Appreciation
    courteous behavior, 6
    overview, 2–3
Aprons, 157
Asphalt driveways, 349–351
Asphalt shingles, 179–181, 185–186
Atoms, 72–77
Attitude
    components, 2–8
    overview, 1–2
Augers, 229, 240
Automatic drain method, 340–341

B

Back miter joints, 149, 155
Bacteria, 314
Baking soda, 237, 240
Ballasts, 122
Band saws, 61–62
Base caps, 155
Base moldings, 154–155
Basin, tub, and tile cleaners, 46
Bathrooms
    cleaners, 46
    clogged drains/toilets, 237–241
    leaks and repairs, 251–252
    public use, 348
    receptacle devices, 94
    sealants, 45
Batteries
    smoke alarms, 109
    uninterruptible power supply, 88
Beading bits, 59
Bees, 328–329
Belly decks, 335–336
Belts, furnace, 260–264
Belt sanders, 59
Beveled blades, 57
Bifold doors, 152–153
Bits, router, 58–59
Black China bristle brushes, 205
Black steel pipes, 231, 232
Blown fuses, 122–123
Blow-out method, 341–342
Bolts, 43
Boring tools, 49–50, 166
Box-end wrenches, 55
Boxes
    fixture installation, 115
    installation, 90–92, 95–98
    receptacle replacement, 113–114
    replacement, 121
    types, 89
Box nails, 42
Brads, 42
Brushes, 205
Bugle screw head, 43
Building codes, 41
Burners, 289–290
Bushes, 338
Butt joints, 149
Butyl rubber, 45
Butyl rubber caulk, 242
Bypass doors, 150–151

C

Cabinet installation, 161–164
Cad cells, 272
Caps, 233
Cap screws, 43
Carbon steel bits, 50
Carpenter's square, 47–48
Carpenter's wood glue, 44
Carpentry
    asphalt shingle installation, 179–181
    base molding installation, 154–155
    cabinet installation, 161–164
    door casing installation, 143–144
    drywall estimations, 131
    engineered panels, 130–131
    grid ceiling system installation, 135–140
    gutter installation, 177–178
    gypsum board cutting/fitting, 168
    hardwood versus softwood, 127–128
    insulation installation, 174–175
    interior door installation, 145–153
    job sheets, 201
    lockset installation, 165–168
    siding installation, 187–199
    wall framing components, 131–134
    wall molding installation, 141–142
    window installation, 176
    wood flooring installation, 159–160
    wood moisture, 128–129
Carriage bolts, 43
Casing nails, 42
Casings
    doors, 143–144
    windows, 156–158
Cast iron pipes, 231, 232
Caulk
    application process, 45–46
    definition, 44
    paint preparation, 207
    plumbing applications, 240–242
    types, 44–45
Ceiling
    drywall materials estimates, 131
    electrical box installation, 90–92
    grid system installation, 135–140
    lighting fixture installation, 116–120
    painting procedure, 211
    paint types, 204
Ceiling fan boxes, 91
Ceiling flats, 204
Ceiling-mounted lighting fixtures, 99
Centrifugal force, 72–73, 74
Chalkline reels, 48, 49
Chandelier-type lighting fixtures, 99, 118
Channel-lock pliers, 54
Charges, 72–77
Chemicals
    reaction prevention, 31
    safety procedures, 21–22
Chilled water systems, 275
Chisels, 149, 166, 349
Chlorinated polyvinyl chloride pipe. *See* CPVC pipe
Chlorine, 344
Choke, 347
Chop saws, 61
Chords, 22–23
Circuit breakers, 123

Circular saws, 56
Clamp-on ammeters, 101, 105, 106
Clamps, 234
Claw hammers, 53
Clean Air Act, 275
Clogged drains/toilets, 237–241
Closed circuits, 76
Closet augers, 229
Clothes dryers, 93, 306–307
Clothing
    chemical hazard procedures, 22
    cooking safety, 31
    electrical safety, 86
    protective equipment, 27–28
    spray painting, 211, 212
Clutch head screws, 43
Cockroaches, 326–327
Coils, 276–277
Cold patch, 349, 350–351
Color, wire, 88–89
Combination squares, 48
Combustion analysis test, 269
Combustion chambers, 271
Common nails, 42
Common slip-joint pliers, 54
Communication skills, 14–15
Competence, 2
Complete circuits, 76
Compound, 209, 210
Compressed air sprayers, 330
Compression faucets, 251–252
Concrete and mortar repair caulk, 45
Concrete walls, 135
Condensate systems, 279–280
Conductors
    grounded circuits, 76
    NM cables, 88–89
    overview, 74–75
    receptacles, 93
    single-phase versus three-phase AC, 87–88
Confidence, 2
Contact cements, 44
Contact poisons, 326
Continuity testers, 100–101
Continuous load, 79
Conventional current flow theory, 75–76
Conversion tables, 355–356
Cooking safety, 31
Cooktops. *See* Stoves
Cooling systems. *See* HVAC systems
Coping saws, 52
Copper tubing
    cutting tools, 229
    measuring and cutting guidelines, 235–236
    soldering guidelines, 242, 244–247
    stove replacement, 291
    types, 231, 232
Cords
    receptacles, 93
    safety tips, 22–23
Core aerators, 343
Corner braces, 133
Corner posts, 132
Coulomb, 75

Couplings, 233
Courtesy, 6, 7
CPVC (chlorinated polyvinyl chloride) pipe, 230, 231
Crosscut saws, 51
Cross-linked polyethylene pipe. *See* PEX pipe
Cross tees, 138, 139
Current
    AC versus DC, 78, 87
    definition, 87
    Ohm's law, 77
Customary measures, 355–356
Customer service
    attitude, 1–8
    checklist, 11
    overview, 1
Cut-in boxes, 89–92
Cutting tools, 49, 50–52
Cylindrical locksets, 165–167

D

Decking, 344, 349
Deck screws, 42
Degreasers, 314
Department of Transportation (DOT), 21–22
Detail sanders, 59
Direct current (DC), 78, 87
Disc sanders, 59
Dishonesty, 4
Dishwashers, 299–301, 311
Door gaskets, 298
Door jacks, 148
Door pulls, 150
Doors
    casing installation, 143–144
    elevator repairs, 319–320
    interior door installation, 145–153
    lockset installation, 165–167
    pest prevention, 325
Door stops, 149
DOT. *See* Department of Transportation
Double-edge wood scrapers, 204
Double-pole switches, 92
Dovetail bits, 58
Drains, clogged, 237–240
Drawer lock bits, 58
Drill bits, 49–50
Drill presses, 62–63
Dropped ceiling, 119–120
Dryers, 93, 306–307
Drywall
    estimating needed materials, 131
    paint preparation, 207, 208–209
    screws, 42
Dumpsters, 315
Duplex nails, 42
Duplex receptacles
    installation procedures, 95–98
    overview, 93, 94
    replacement, 113–114

E

Edge bits, 59
Edging, 336–337
Eggshell paints, 204

8-inch adjustable wrenches, 229
Electrical boxes. *See* Boxes
Electrical installations
    devices and fixtures, 92–100, 114–120, 123
    equipment and supplies, 88–92
    measurement instruments, 100–106
    overview, 85
    troubleshooting, 125
Electrical maintenance
    equipment replacements, 109–113, 121–123
    HVAC systems, 272–274
    job sheet, 125
    regular tests, 108–109
    troubleshooting process, 107–108
Electrical safety
    installations, 86–87
    ladder procedures, 25
    power tools and cords, 22–23
    troubleshooting process, 107
Electrician's pliers, 54
Electricity
    AC versus DC currents, 78
    backup systems, 80
    basic theory, 71–77, 87
    electrical load calculations, 77–78
    emergency circuits, 79–80
    load carry calculations, 79
    purpose, 71
    wire size measures, 79
Electric range receptacles, 93
Electric spin drain cleaners, 229
Electric stoves, 293–298, 309
Electric water heaters, 248–249, 250
Electromotive force (EMF), 77
Electrons, 72–77, 78
Electron theory, 75
Elements, 72, 73
Elevators, 319–323
11-in-1 multipurpose tools, 204
Emergency backup systems, 80, 88
Emergency circuits, 79–80
Emery sandpaper, 60
EMF. *See* Electromotive force
Empathy, 3–4
Enamels, 205
Engineered panels, 130
Engine repair, 346–348, 357–358
English measurement system, 355–356
Environmental Protection Agency (EPA), 21–22
Epoxy, 44
Estate sprayers, 330, 331
Evaporator coils, 276–277
Evergreen bushes, 338
Exhaust fans, 302
Explosions, 22
Extension ladders, 25
Eye protection, 27

F

Faceplate markers, 166
Fall protection
    job sheet, 37
    ladder procedures, 23–26
    overview, 20–21

Fan-rated ceiling boxes, 91
Fasteners, 41–46, 65
Faucets, 251–252
Fiberboards, 130
Filters
    coil cleaning, 276
    replacement, 266–267, 269–270
Finger joint bits, 58
Finishing nails, 42
Finishing tools, 204–206, 219
Fire
    classes, 29–31
    electrical safety, 22
    prevention tips, 31
Fire alarms, 108–109
First impressions, 2
Fittings, 232–233
Fixed routers, 57
Flapper assembly, 251
Flat brushes, 205
Flat paints, 204
Flat screw head, 43
Flexible insulation, 174–175
Flies, 328
Flint sandpaper, 60
Floor guides, 151
Flooring, wood, 159–160
Flowering shrubs, 338
Flue pipes, 270
Fluorescent bulbs, 94, 121–122
Fluorescent lighting fixtures, 116, 119
Flush bits, 59
Flush cut blades, 57
Flush pulls, 150
Flux, 244
Flywheel keys, 348
Footwear, 25
45-degree elbow fittings, 233
4" boxes, 91
14-inch pipe wrenches, 228
Four-way switches, 92, 93
Framing process
    components, 131–134
    grid ceiling installation, 135–140
Framing squares, 47–48
Frustration, 4
Fume hoods, 21
Fumigants, 326
Furnace
    electrical repairs, 272–274
    gas unit maintenance, 267–268
    general maintenance steps, 260–266
    hot water/steam boilers, 268–269
    job sheets, 283, 285, 287
    oil burner/boiler maintenance, 269–272
    standing pilots, 274–275
Furniture, 207
Fuses, 122–123

G

Galvanized pipes, 231
Galvanized screws, 43
Garbage containers, 326, 349
Garnet sandpaper, 60
Gas furnaces, 267–268
Gas stoves, 289–293
Gas tanks, 346
Gas water heaters, 249–251
Generators, 79, 80
GFCI. See Ground-fault circuit interrupters
Gloss paints, 204
Gloves, 27, 314, 315
Glue, 44, 46
Grass shears, 336, 337
Grease fittings, 278
Greeks, 71–72
Green lumber, 128–129
Grit range, 60
Grooving bits, 58
Grounded circuits, 76
Ground fault circuit interrupters (GFCI)
    box installation, 98
    electrical safety, 22
    overview, 79
    receptacles, 93, 94
    replacement, 113–114
    tests, 109
Groundskeeping
    asphalt driveway repairs, 349–351
    bushes and shrubs, 338
    edging, 336–337
    irrigation systems, 338–343
    job sheets, 353
    lawn aeration, 343
    mowing, 335–336
    mulching, 337
    overview, 335
    pool maintenance, 344
    public area maintenance, 348–349
    small engine repair, 346–348, 357–358
    snow plowing, 345
Guardrails, 20
Guide pins, 152
Gutters, 45, 177–178
Gypsum board, 131, 168

H

Hacksaws, 51, 228, 236
Hallways, 348
Hammers, 52–53
Hand aerators, 343
Handheld edgers, 336
Hand sprayers, 329–330
Hand tools
    definition, 46
    electrical safety, 87
    job sheet, 69
    types, 46–55
Hanger lags, 136
Hanger straps, 234
Hanger wires, 136, 137
Hardboard, 131
Hardwood, 127–128, 130
Harnesses, 20, 28
Hazards
    chemical substance procedures, 21–22
    electrical procedures, 22–23
    fall protection, 20–21
    fire prevention, 31
    ladder procedures, 23–26
    lifting procedures, 28–29
    OSHA creation and purpose, 19–20
    power tool procedures, 22–23
    protective equipment, 27–28
    safety job sheet, 33, 35, 37, 39
    work tasks assignments, 14
Header casings, 158
Headers, 131
Heat exchangers, 269
Heating, ventilation, and air-conditioning (HVAC) systems. See HVAC systems
Heating elements, 294
Helmets, 27, 28
Hex screw head, 43
High-density fiberboard, 131
High-gloss paints, 204
High speed steel (HSS) bits, 50
Hinges, 147, 148, 149
Hole saws, 50
Honesty, 4–5
Hooks, 46–47
Horizontal siding
    installation, 187–190
    vinyl application, 194–198
Hornets, 328
Hose-end sprayers, 329, 330
Hot water/steam boilers, 268–269
HSS bits. See High speed steel bits
HVAC systems. See also Air conditioners
    chilled water systems, 275
    condensate systems, 279–280
    electrical device repairs, 272–274
    evaporator coils, 276–277
    filter replacement, 266–267
    gas-fired furnace heat source, 267–268
    general furnace maintenance, 260–266
    hot water/steam boiler maintenance, 268–269
    job sheets, 283, 285, 287
    motor lubrication, 277–278
    oil burner/boiler maintenance, 269–272
    overview, 259
    pest prevention, 326
    standing pilots, 274–275
    through-the-wall systems, 280–281
    troubleshooting system, 278–279

I

Ice choppers, 345
Ice makers, 298
Incandescent bulbs, 94
In-line ammeters, 101, 105
Inorganic mulches, 337
Instructional information, 14–15
Insulation, 174–175
Interior doors
    casing application, 143–144
    installation, 145–153
Interlock safety devices, 315
Irrigation systems, 338–343

J

Jamb extensions, 158
J-channels, 194–199

Joinery bits, 58
Joint compound, 209, 210

K
Key hole bits, 59
Kitchens
    clogged drains, 237
    cooking safety, 31
    receptacle devices, 94
    sealants, 45

L
Ladders
    job sheet, 39
    safety procedures, 23–26
    smoke alarm replacement, 110
Lag screws, 42
Laminate routers, 57, 58
Landscaping, 338
Language, 6, 7–8
Lanyards, 20
Latex paints, 204, 206, 214
Lawn aeration, 343
Law of centrifugal force, 72–73
Law of charges, 72, 73, 74
Levels, 48–49, 229
Lifting procedures, 28–29
Light bulbs, 94, 121–122
Lighting fixtures
    installation, 115–120
    overview, 94, 99–100
    pest prevention, 326
    replacement, 121
Liquid drain cleaners, 237–238
Listening, 3, 7
Lobby areas, 349
Locking tape measures, 228
Lock-out/tag-out procedures, 86
Locksets, 165–168
Low-density fiberboard, 130
Lugs, 198

M
Machine screws, 42
Magnets, 72
Manual drain method, 338–339
Manufactured cabinets, 161–164
Manufacturer's labels, 21
Masonite, 131
Masonry bits, 50
Masonry nails, 42
Mason's hammers, 52
Matches, 31
Material safety data sheets (MSDSs), 21
Matter, 72
Measuring tools, 46–49. *See also specific tools*
Medical alert systems, 109
Medium-density fiberboard (MDF), 130
Megohmmeter, 101
Messages, 5
Metal pipes, 231
Metric measures, 355–356
Microwaves, 303
Midget copper cutters, 229
Miter boxes, 230
Miter saws, 60–61

Molding
    base trim, 154–155
    walls, 141–142
    windows, 156–158
Mortar repair caulk, 45
Mosquitoes, 328
Motor lubrication, 277–278
Mowing, 335–336, 353
MSDSs. *See* Material safety data sheets
Mulching, 337
Multispur bits, 166

N
Nails, 42, 207, 208
Nap, 206
National Electrical Code (NEC), 79, 86
Needle nose pliers, 54
Negative self-talk, 3
Neutrons, 72–77
NFPA 70 standard, 86
90-degree elbow fittings, 233
NM cable, 88–89
Nucleus, 74
Nuts, 43
Nylon brushes, 205

O
Occupational Safety and Health Act (OSHA)
    electrical safety, 86
    overview, 19
    purpose, 20
Ohmmeters, 101, 105
Ohm's law, 77–78, 83
Oil-based paint brushes, 205, 213
Oil-based paints, 204, 206, 214
Oil burner/boiler maintenance, 269–272
Oil changes, 346
Oil filters, 269–270
Oil ports, 277–278
Old-work electrical boxes, 89–90
Omega symbol, 77
100% silicone caulk, 44–45
100% silicone kitchen and bath sealant, 45
120-volt receptacle, 92–93, 121
Open-end wrenches, 55
Organic mulches, 337
Oriented strand board (OSB), 130
OSHA. *See* Occupational Safety and Health Act
Oval brushes, 205
Oval screw head, 43
Ovens
    electric, 296–297
    gas, 290–291

P
Paint brushes, 205, 213, 223
Painter's caulk, 44
Painting
    benefits, 203
    cleaning guidelines, 212–214
    job sheets, 217, 219, 221, 223, 225
    paint selection, 204
    steps, 210–212

    surface preparation, 203, 206–210
    tools, 204–206
Paint rollers
    cleaning guidelines, 212–213
    covers, 206
    frames, 206
    job sheets, 223
    painting guidelines, 210–211
Paint scrapers, 204
Paint sprayers
    cleaning tips, 213
    job sheets, 223
    steps for use, 211–212
Pancake boxes, 91
Paneling, 169–173
Panel siding, 192
Pan screw head, 43
Parking lots, 345, 349–351
Particleboard, 130
Particleboard screws, 42
Partition intersections, 133
Patios, 349
Pendant-type incandescent lighting fixtures, 99
PE (polyethylene) pipe, 231
Pesticides, 326–331
Pest prevention
    job sheets, 333
    overview, 325
    pesticides, 326–331
    steps, 325–326
PEX (cross-linked polyethylene) pipe, 231
Phillips screwdrivers, 53
Phillips screws, 43
PH levels, 344
Pilot holes, 290
Pilot lights, 274–275
Pipe staples, 235
Pipe wrenches, 228
Piping
    assembly, 242–247
    cutting tools, 229, 230
    fittings, 232–233
    measuring and cutting guidelines, 235–236
    support and hangers, 233–235
    types, 230–232
Pivot sockets and pins, 152, 153
Planes, 57
Plants, 338
Plastic laminates, 131
Plastic pipes
    cutters, 230
    cutting tools, 230
    joining guidelines, 243
    types, 230–231
Plates, 131
Pliers, 54
Plies, 130
Plugs, 233
Plumb bobs, 49
Plumber's putty, 242
Plumbing
    caulk, 240–242
    clogged drains/toilets, 237–240
    codes, 248
    importance, 228

job sheets, 255, 257
leak repairs, 251–252
putty, 242
tools and equipment, 228–235
water heater adjustments/
   replacement, 248–251
Plunge routers, 57
Plungers, 229, 237, 239
Plywood, 130
Polyester brushes, 205
Polyethylene pipe. *See* PE pipe
Poly/nylon brushes, 205
Polyurethanes, 205
Polyvinyl chloride pipe. *See* PVC pipe
Pool maintenance, 344
Portable power tools, 56–60
Posidriv® screws, 43
Positive self-talk, 3
Power, 87
Power backpack sprayers, 330, 331
Power outages, 80
Power sanders, 204
Power saws, 56
Power switch, 23
Power tools
   electrical safety, 87
   tips for use, 22–23
   types, 56–63
Pressure washers, 204
Pressurized cans, 329, 330
Primer, 205, 211
Priorities, 13–14
Progress reports, 15
Propane torch, 242, 244–247
Protectants, 326
Protective equipment, 27–28
Protons, 72–77
Pulleys, 260, 261, 264–266
Pull knobs, 153
Push mowers, 335, 336
Push-pull hand pump sprayers, 329, 330
Putty knife, 204, 205
PVC (polyvinyl chloride) pipe, 230, 231

Q
Quick connector plugs, 110

R
Racking, 160
Raised panel bits, 59
Random orbit sanders, 59
Range hoods, 301–303
Ranges. *See* Stoves
Reamers, 230
Receptacles
   installation, 92–94
   paint preparation, 207
   replacement, 112–114
Recessed-can incandescent lighting
   fixtures, 99
Reciprocating saws, 56
Rectangular flush pulls, 150
Recycling, 214
Reducers, 233
Refrigeration circuits, 275
Refrigerators, 298
Rekeying locks, 167

Reliability, 5–6, 14
Reset buttons, 94, 109
Resin, 204
Resistance, 77, 78, 87
Responsiveness, 6, 7–8
Ribbons, 133
Ridge caps, 181
Riding mowers, 335–336
Rile and stile bits, 58
Ripsaws, 51
Robertson screwdrivers, 53
Robertson type screws, 43
Rodents, 327
Roller covers, 206
Roller frames, 206
Roller hangers, 150, 151
Roll roofing, 182–185
Romex, 88–89
Roofing
   asphalt shingle installation,
      179–181, 185–186
   nails, 42
   repair caulk, 45
   roll roofing installation, 185–185
Rough sills, 132
Round ceiling boxes, 91–92
Round-nose bits, 58
Round-over bits, 59
Round screw head, 43
Routers
   bit types, 58–59
   overview, 57
   tips for use, 58
   types, 57

S
Saber saws, 56
Safety
   chemical substance procedures,
      21–22
   electrical procedures, 22–23,
      86–87, 107
   fall protection, 20–21
   fire prevention, 31
   job sheet, 33, 35, 37, 39
   ladder procedures, 23–26
   lifting procedures, 28–29
   OSHA creation and purpose, 19–20
   paint preparation, 207
   pesticides, 331
   power tools, 22–23
   protective equipment, 27–28
   soldering, 246
   stationary tools, 63
   work task assignments, 14
Safety switches, 23
Sanders, 59–60, 204
Sandpaper, 60
Satin paints, 204
Saws
   hand tools, 50–52
   power tools, 56
Scrapers, 204
Screwdrivers, 53–54
Screw eyes, 136, 137
Screws, 42–43, 208
Self-talk, 2–3

Semi-gloss paints, 204
Shallow boxes, 91
Sheet metal screws, 42
Sheet paneling, 169–171
Sheetrock, 90
Shingles
   asphalt, 179–181, 185–186
   wooden, 192–193
Shoe molding, 155
Short circuits, 76
Shower seals, 251
Shrubs, 338
Side-cutting pliers, 54
Sidewalks, 345
Siding, 187–199
Sight glass, 270
Silicone carbide sandpaper, 60
Silicone caulk and sealant, 44, 45, 241
Siliconized latex caulk, 44
Sincerity, 4, 6
Single-gang nail-up boxes, 89–90
Single-phase AC, 87–88
Single-pole switches, 92
Single receptacle, 93
Sinks
   clogs, 237, 238, 239
   repairs, 251–252
Sink snakes, 238, 239
6-inch adjustable wrenches, 229
Sledge hammers, 52
Slotted screwdrivers, 53
Slotted screws, 43
Small engine repair, 346–348, 357–358
Small motorized sprayers, 330–331
Smoke alarms
   replacement, 110–111
   tests, 108–109
Smoking, 31
Snaplock punches, 197
Snow plowing, 345
Sockets, 152, 153
Socket wrenches, 55
Soffits, 198
Softboard, 130
Softwood, 127–128
Soldering pipe, 242–247
Solid wood paneling, 172–173
Solvents, 46
Spade bits, 50
Spark ignition ranges, 290
Spark plugs, 346, 348
Spiders, 327
Spills, 22
Spindle sanders, 61
Spot removers, 46
Sprayers, 329
Spring irrigation startup, 342–343
Sputtering engines, 347–348
Squares, 47–48
Square screwdrivers, 53
Stains, 205
Stairways, 345, 348
Standing pilots, 274–275
Staples, 42
Stationary tools, 60–63
Step flashing method, 1⁰
Step ladders, 23–26

Stomach poisons, 326
Stools, 156, 157
Stove bolts, 43
Stoves
    electric stove repair/replacement, 293–298
    gas stove repair/replacement, 289–293
    job sheets, 309
Straight bits, 58
Straight ladders, 23–26
Strainers, 211
Straps, 117, 118
Stress, 2, 6
Strike plates, 146
Striker plates, 166–167
String trimmers, 358
Studs, 131
Subfloors, 159
Surface-mount fluorescent fixtures, 94, 99
Surface-mount incandescent fixtures, 94, 99
Switches
    overview, 92
    paint preparation, 207
    replacement, 112
Systemics, 326

T

Table saws, 62
Tape measures, 46–47, 228
Tee fittings, 233
Tension gauges, 262, 263
Termites, 329
Test equipment, 87
Thinner, paint, 211
Three-phase AC, 87–88
Three-way switches, 92, 93
Through-the-wall air conditioners, 280–281
Ticks, 328, 329
Tile caulk, 45, 240–242
Tile cleaners, 46
Toilets
    clogs, 237–238, 240–241
    repairs, 251
Tone of voice, 7
Toolboxes, 230
Tools. *See also specific tools*
    job sheet, 65, 67, 69
    safety, 63, 87
    selection, 41
Torpedo levels, 229
Torx screwdrivers, 53
Toxic hazards. *See* Hazards
Transformers, 78
Transformer springs, 272
Trash compactors, 313–317
Trigger pump sprayers, 329, 330
Trim
    baseboards, 154–155
    walls, 141–142
    windows, 156–158
Trimmers, 132
Trip/reset buttons, 94, 109
T-slot bits, 59
Tubes, 94
Tubing cutters, 235, 236
Tubs
    caulk, 45, 46, 240–242
    cleaners, 46
    clogs, 239–240
Twinfast drywall screws, 42
Twisted bits, 50
240-volt receptacle, 92, 93
Type K copper tubing, 231, 232
Type L copper tubing, 231, 232
Type M copper tubing, 231, 232

U

U-bolts, 234
Undersill trim, 198
Uninterruptible power supply (UPS), 88
Unions, 233
U-shaped frames, 206

V

VA. *See* Volt-amps
Valence shell, 74
Valve guides, 347
Vented ridge caps, 181
Ventilation systems. *See* HVAC systems
Vents, 306–307
Vertical tongue-and-groove siding, 190–191
V-groove bits, 58
Vinyl latex caulk, 241
Vinyl siding, 194–199
Vise grip pliers, 54
Voltage
    AC versus DC current, 78
    definition, 77, 87
    Ohm's law, 77–78
    single-phase versus three-phase AC, 87–88
    testers, 101, 102–104, 314
    troubleshooting process, 107
Volt-amps (VA), 87
Voltmeters, 101
Volts, 87

W

Wait times, 4–5
Wallboard saws, 52
Wall-mounted lighting fixtures, 99
Walls
    base molding installation, 154–155
    drywall materials estimates, 131
    framing components, 131–134
    grid ceiling installation, 135–140
    molding installation, 141–142
    painting procedure, 211
    siding installation, 187–199
Warning line systems, 20
Washerless faucets, 251–252
Washers, 303–306
Wasps, 328
Water-based paint brushes, 205, 213
Water-based paints, 214
Water heaters, 248–251
Water inlet valves, 304–305
Wattage, 77, 78, 87
Weatherproof boxes, 91–92
White China bristle brushes, 205
White spirit, 213
Windows
    installation process, 176
    paint preparation, 207
    pest prevention, 325
    trim installation, 156–158
Winterizing irrigation systems, 338–342
Wire brushes, 213
Wire hangers, 235
Wire screening, 209
Wiring
    emergency circuits, 80
    gauge/current chart, 357
    HVAC repairs, 273–274
    NEC guidelines, 79
    smoke detector replacement, 110
    types, 88–89
Wood
    flooring installation, 159–160
    glue, 44
    hardwood versus softwood, 127–128
    job sheets, 201
    moisture content, 128–129
    paneling, 172–173
    screws, 42
    shingles and shakes, 192–193
Work tasks
    assignments, 14–15, 17
    completion documentation, 15
    overview, 13
    priorities, 13–14
Woven valley method, 185
Wrenches, 55, 228, 229